SpringerBriefs in Molecular Science

History of Chemistry

Series Editor

Seth C. Rasmussen, Department of Chemistry and Biochemistry, North Dakota
State University, Fargo, ND, USA

Springer Briefs in Molecular Science: History of Chemistry presents concise summaries of historical topics covering all aspects of chemistry, alchemy, and chemical technology. The aim of the series is to provide volumes that are of broad interest to the chemical community, while still retaining a high level of historical scholarship such that they are of interest to both chemists and science historians.

Featuring compact volumes of 50 to 125 pages, the series acts as a venue between articles published in the historical journals and full historical monographs or books.

Typical topics might include:

- An overview or review of an important historical topic of broad interest
- Biographies of prominent scientists, alchemists, or chemical practitioners
- New historical research of interest to the chemical community

Briefs allow authors to present their ideas and readers to absorb them with minimal time investment. Briefs are published as part of Springer's eBook collection, with millions of users worldwide. In addition, Briefs are available for individual print and electronic purchase. Briefs are characterized by fast, global electronic dissemination, standard publishing contracts, easy-to-use manuscript preparation and formatting guidelines, and expedited production schedules. Both solicited and unsolicited manuscripts are considered for publication in this series.

More information about this subseries at http://www.springer.com/series/10127

Gary Patterson

Chemistry in 17th-Century New England

 Springer

Gary Patterson
Department of Chemistry
Carnegie Mellon University
Pittsburgh, PA, USA

ISSN 2191-5407   ISSN 2191-5415   (electronic)
SpringerBriefs in Molecular Science
ISSN 2212-991X
SpringerBriefs in History of Chemistry
ISBN 978-3-030-43260-7   ISBN 978-3-030-43261-4   (eBook)
https://doi.org/10.1007/978-3-030-43261-4

This Springer imprint is published by the registered company Springer Nature Switzerland AG
The registered company address is: Gewerbestrasse 11, 6330 Cham, Switzerland

Acknowledgements

I would like to thank the History of Chemistry Division of the American Chemical Society for its support of a symposium on chemistry in America before 1876. I would also like to thank my co-organizers, Ed Cook and Seth Rasmussen. The primary source research for this work has benefitted from the extensive collection of Winthrop Family papers by the Massachusetts Historical Society. I would like to thank Trevor Levere, Louise Palmer and Seth Rasmussen for careful reading and good suggestions for improvement.

Contents

Chapter 1
Introduction

Abstract The world of 17th century New England is introduced. The five worlds of John Winthrop, the Younger, are introduced: Political, Religious, Scientific, Industrial and Social. The key characters in this fascinating drama are identified.

Chemistry was being practiced in New England long before the Puritans arrived in Massachusetts to create a "commonwealth in the wilderness [1]." The focus of this book will be the chemistry associated with the developing Massachusetts Bay and Connecticut Colonies. There were no universities to welcome the Puritans, and the founding of Harvard College in 1637 was not intended to have any industrial consequences. Nevertheless, a vibrant world of chemistry did grow in New England. The story of this chapter in American chemistry has received a major boost since the recovery of the Winthrop papers [2]. The Massachusetts Historical Society has collected the papers and correspondence of many generations of Winthrops. John Winthrop, Sr., was the first Governor of the Massachusetts Bay Company; John Winthrop, Jr., was the Governor of Connecticut; both sons of John Winthrop, Jr., became Governors in New England, and many famous Winthrops are buried in the Kings Church graveyard in Boston. The present book is hardly the first to tell this story, and I hope it is not the last, but now is an excellent time to advance our understanding of the role of chemistry in the development of the American experience.

In the beginning of the seventeenth century, chemists divided all material things into three parts: animal, vegetable and mineral [3]. The task of chemistry was to take what was bountifully provided by Providence and modify it for the use of humankind. New England could also be divided into three realms: sky, land and sea. It was believed by the Puritans that all their needs would be met if they applied themselves diligently to the task of founding a commonwealth [4]. This vision was reinforced by Puritan divines, but it was largely the creation of men like John Smith [5] (1579–1632, Fig. 1.1) and John Winthrop, Sr., (1588–1649, Fig. 1.2).

The Massachusetts Bay Company was a commercial venture [6]. While there were religious aspects to an explicitly Puritan commonwealth, the proprietors in England expected to make money. How were they going to do that? The king expected the colonies to be a source of wealth for him. London merchants expected to sell finished goods to the needy colonists. New England had substantial natural resources, but in

© The Author(s), under exclusive license to Springer Nature Switzerland AG 2020

G. Patterson, *Chemistry in 17th-Century New England*,
SpringerBriefs in History of Chemistry,
https://doi.org/10.1007/978-3-030-43261-4_1

order to reach a market for these goods, they needed to be transported somewhere else. This increased the costs and reduced any profit.

The most obvious natural resource in New England was trees [1]. Eventually, local craftsmen used them to build both their own homes and sailing ships. Americans could now sail their own ships and could engage in triangular trade. With a large fleet, Americans could exploit the next great natural resource, the sea. Fish existed in abundance, but in order to ship them to England, a great amount of salt was needed. Salt water was readily available and Americans devised methods for processing sea water into usable salt. It no longer needed to be purchased from England. In order to exploit the rocky soil, good plows were needed. Eventually, Americans developed a local iron industry. They had plenty of iron ore, wood for charcoal, and free air to process everything. As America became more and more independent of England for the necessities of life, the crown turned to taxation as a way to extract value from the colonies. But, collection proved to be too expensive!

Fig. 1.2 John Winthrop, Sr.
(1588–1649). Courtesy
American Antiquarian
Society

1.1   John Winthrop, Jr.

The story of how the Americans developed their own chemical enterprises in New England centers on one person, John Winthrop, Jr., (1606–1676, Fig. 1.3), often referred to as John Winthrop, The Younger. While he followed in his father's footsteps in Massachusetts by obtaining a Royal Charter for Connecticut, and hence played a dominant political role in New England, he also established local industries that freed the colonists from dependence on the mother country and made possible a true commonwealth in the wilderness. This book will follow the life of John Winthrop, Jr., and will introduce the other actors in this seventeenth century drama. It is a great story and every historian of chemistry should know it.

The study of the life of John Winthrop, Jr., has been made much easier by the publication of three books: (1) *A sketch of the life of John Winthrop, the younger, founder of Ipswich, Massachusetts, in 1633,* by Thomas Franklin Waters and Robert C. Winthrop [7]. It first appeared in 1899 as a contribution from the Ipswich Historical Society; (2) *The Younger John Winthrop* by Robert C. Black, III, a historian from Trinity College, Connecticut, which appeared in 1966 [8]; and (3) *Prospero's America: John Winthrop, Jr. Alchemy, and the Creation of New England Culture, 1606–1676,* by Walter W. Woodward, the Connecticut State Historian and a Professor at the University of Connecticut, which appeared in 2010 [2]. In addition, primary

Fig. 1.3 John Winthrop, the younger (1606–1676) (Harvard University portrait collection, Gift of Robert Winthrop, representing the Winthrop family, to Harvard University, 1964)

source research can be carried out at the Massachusetts Historical Society where the *Winthrop Papers* are archived.

The study of the history of New England has progressed from Colonial myths to a well-nuanced history in the last 50 years. The Dean of Massachusetts historians was Perry Miller (1905–1963) of Harvard University. One of his many books is: *The New England Mind: the Seventeenth Century* (1939) [9]. Another seminal volume from Harvard was published by Samuel Eliot Morison (1887–1976): *Builders of the Bay Colony* (1930) [10]. It contains a nice chapter on John Winthrop, Jr., and chapters on many of the key figures in the founding of Massachusetts.

The story of John Winthrop, the Younger, is not a simple one, and it is easy to get lost in the details. One organizing principle is to construct a map of his many worlds of discourse and social concourse. One world was his religious world. He was a Puritan, but he was probably the most irenic Puritan in New England [11]. He remained on good terms with his Puritan father, and with the Puritan grandees of England. He was also on intimate terms with Roger Williams, who was banished from Massachusetts and founded Rhode Island. He argued in favor of allowing Quakers freedom of religion in New England. His close friends included Roman Catholics, like Sir Kenelm Digby, the privateer and alchemist. Another of his closest friends, Edward Howes, was fascinated by the Rosicrucian movement. But, of all the religious figures in his world, Samuel Hartlib and John Amos Comenius, the Moravian bishop, had the greatest influence. John Winthrop hoped to help inaugurate the Great Instauration

[12] and bring true Pansophism [13] to America. Perhaps he infected New England with enough religious tolerance to allow the formation of the United Colonies.

Another obvious world for John Winthrop, Jr., was his political world. After all, he served as the Governor of Connecticut for almost 20 years. One of his primary colleagues in his political world was his father, John Winthrop, Sr. Of even more importance for the formation of his political character was his uncle, Emmanuel Downing. Roger Williams, as Governor of Rhode Island, had many political dealings with John Winthrop, Jr. A voyage from England to Massachusetts with Henry Vane, the Younger, helped John Winthrop to understand the workings of English government, and later Vane aided the Colonists from his position in the Parliamentarian government. When in England, John Winthrop, Jr., was busy learning from the Puritan aristocracy. Three notable figures were Viscount Say and Sele, Lord Brooke and the Earl of Lincoln. After the Restoration, it was time to obtain a Royal Charter for Connecticut. Sir Robert Moray was close enough to the King to introduce Winthrop to all the right people. When needed, Robert Boyle could be counted on to plead the case for New England. John Winthrop, Jr., even reached the point where he could correspond with the Lord Chancellor Clarendon. In a monarchical system, nothing happens without powerful friends. But, New England was run using a different sort of politics. John Winthrop, Jr., knew how to play in both worlds, and to use one to gain advantage in the other.

The largest world for John Winthrop, the Younger, was his scientific world. He fell in love with alchemy while working with Edward Howes. He purchased one of the greatest alchemical libraries of the seventeenth century [14] and his books were his constant companions. Francis Kirby also sold him many books and chemicals. Another of his friends and suppliers was Robert Child. Many of his scientific friends were associated with the Circle of Samuel Hartlib. These included Sir Kenelm Digby, Sir Robert Moray FRS and Robert Boyle FRS. In Europe, the Circle centered on the brothers Abraham and Johann Kuffler. Their close associates included Johann Glauber, Johannes Tanckmarus and Johann Moraien. John Winthrop, Jr., was also a founding member of the Royal Society. When he died, he was memorialized by Henry Oldenburg, the Secretary of the Royal Society: "a person most curious and able, and of a nature prone to pardon".

But, rather than being merely a scholar, John Winthrop, Jr., had an industrial and commercial world. The Massachusetts Bay Company was a commercial venture and Winthrop served it well. He organized the second voyage of settlers to Massachusetts. He was elected as one of the Assistants of the Corporation. He founded the town of Ipswich. In order to further the fishing industry, he launched a saltworks. After the English Civil war made New England both poor and devoid of many necessary items, he helped to found an ironworks. There were many groups of Puritans that desired to either settle in New England, or at least make some money there. He founded the town and fort at Saybrook, Connecticut. He also founded his own plantation at New London. He founded a livestock enterprise on his own island, Fishers Island. He also founded a saltworks at New Haven. His greatest enterprise was founded as primarily a medical charity: he treated patients throughout New England for free. As a result, he was badly in monetary debt at his death, but was beloved by many people, both friends and official adversaries.

And finally, John Winthrop, the Younger, had a lively personal world. He was apparently one of the most charming characters of the seventeenth century. People from all over the European world loved him. He maintained a warm relationship with his step-mother, the former Margaret Tyndal, throughout her life. His aunt and uncle, Lucy Winthrop and Emmanuel Downing, were both good guardians and adult friends. Edward Howes was a close friend and correspondent throughout his adult life. Another of his longstanding correspondents was the Reverend Henry Jacie (1603–1663). They met during Winthrop's preparation to emigrate from England in 1631 at Jacie's home in Assington, near Groton. Like John Winthrop, Jr., he was irenic in religion and, though a Puritan clergyman, was a champion in advocating for tolerance of the Jews in England. He was a natural philosopher and he and John Winthrop discussed many things in their correspondence. Jacie also kept Winthrop abreast of the news in England, especially agitations against Puritans and New England. After John Winthrop, Jr., had established the new town of Agawam (later Ipswich) and achieved a stable economy and government, it was time to hire a minister to lead the worship and teach the word. He obtained both a good clergyman and a true friend, Nathaniel Ward (1578–1652).

Throughout the history of seventeenth century New England, chemistry played an important part. Importing chemicals from England was expensive and not always possible. Developing local sources of industrial chemicals was essential for long-term survival. There were very few members of the Royal College of Physicians in Colonial New England. But, people got sick and looked to the only educated people available for help; this was usually the parish minister. Puritans believed in an educated clergy, and such pastors believed that they should truly shepherd their flocks. This meant that many of the graduates of Harvard College ended up practicing medicine, at least to some degree [15]. They usually employed some form of Paracelsian iatrochemistry [16]. This required that they manufacture their own medicines. What was almost completely lacking in Colonial New England was any form of academic chemistry. There were no medical schools in Massachusetts, like there were in Germany, France or Italy. The craft was learned by interning with a practicing physician. There was also a large group of aspiring adepts who learned their alchemy from John Winthrop, Jr., and from "books of secrets [17]." They largely achieved bankruptcy for their families. No successful preparations of the "philosopher's stone" were achieved in Colonial New England. Nevertheless, a tradition of alchemy dominated seventeenth century New England among those rich enough or foolish enough to devote substantial time and money to this endeavor.

With such a rich life to examine and with such a vast supply of primary source materials, a great story can be told. The focus will always be on the chemistry of the story, but enough other details will be included to provide a proper context. The story of seventeenth century New England is about much more than witches and Native American massacres.

References

1. Innes S (1995) Creating the commonwealth: the economic culture of puritan New England. Norton, New York
2. Woodward WW (2010) Prospero's America: John Winthrop, Jr., Alchemy, and the creation of New England culture. UNC Press, Chapel Hill
3. Libavius A (1597) Alchemie, Frankfurt
4. Winthrop J (1908) Winthrop's journal: history of New England, 1630–1649. In: Hosmer JK (ed) Scribner, New York
5. Smith J (1907) The general historie of Virginia, New England and the summer isles. Macmillan, New York
6. Rose-Troup F (1930) The Massachusetts Bay company and its predecessors. Clearfield, New York
7. Waters TF, Winthrop RC (1899) A sketch of the life of John Winthrop, the younger, founder of Ipswich, Massachusetts, in 1633. Ipswich Historical Society, New Chum
8. Black RC (1966) The younger John Winthrop. Columbia University Press, New York
9. Miller P (1939) The New England mind: the seventeenth century. The Belknap Press of Harvard University, London
10. Morison SE (1930) Builders of the Bay Colony. Cambridge University Press, Cambridge
11. Lloyd-Jones DM (1987) The puritans: their origins and successors. Banner of Truth Trust, Carlisle, London, Scottsdale
12. Webster C (1975) The great instauration: science, medicine and reform 1626–1660. Duckworth, London
13. Comenius J A (1916) The great didactic of John Amos Comenius. (Trans: Keatinge MW). London
14. Wilkinson RS (1966) The alchemical library of John Winthrop, Jr. Ambix 18:139–186
15. Watson PA (1991) The angelical conjunction: the preacher-physicians of colonial New England. University of Tennessee Press, Knoxville
16. Webster C (2008) Paracelsus: medicine, magic and mission at the end of time. Yale University Press, London
17. Kavey A (2007) Books of secrets: natural philosophy in England, 1550–1600. University of Illinois Press, Urbana

Chapter 2
The Puritan Mind in the Seventeenth Century

Abstract The history and philosophy of Puritanism is presented and related to John Winthrop, Jr. The scientific world of Francis Bacon is also presented. Winthrop was fully committed to both of these programs for human flourishing.

The worldview of the seventeenth century was very different than that of the twenty-first century [1]. Religious questions still burned and wars abounded everywhere. While the English Reformation under Henry VIII had produced a nominally unified national church in England, the realities were very different. Since the Anglican Church was established, every parish was under the control of the central authorities, at least to some extent. The real power at the local level was the Lord of the Manor. He was expected to provide both a living for all his neighbors and to support the local church. If the Lord was actually still a Roman Catholic in spirit, he encouraged the use of high church practices [2].

The broader world of Protestant Christendom was badly fragmented. But, a useful distinction can be made into five categories: Lutheran, Calvinistic (Reformed), Zwinglian (Swiss), Anabaptist and Independent. Within the official Anglican Church, there were some who felt that the English Reformation had not been much of a "re-formation," and that many of the worst aspects of Roman Catholicism still remained in local practice. D. M. Lloyd-Jones traces the beginning of the Puritan movement to William Tyndale (1494–1536) [3]. What made Tyndale a prototype of a Puritan was that he defied the bishops when he translated and printed his English language Bible. In order to do this, he left England without royal permission. Puritans refused to allow "bishops" to enforce a false orthodoxy and were willing to emigrate to follow their faith. Tyndale was arrested in Europe, at the request of the English king, and executed in a cruel and barbaric manner involving both strangulation and burning at the stake. Puritans never forgot!

When Archbishop Thomas Cranmer (1489–1556) tried to move the Anglican Church more in the direction of the Protestant reformation, he met with considerable opposition. He did help produce the Book of Common Prayer and the doctrinal Thirty-Nine Articles, but he believed his highest responsibility was to maintain as much unity as he could. For his trouble he was executed when Mary Tudor became Queen and demanded Roman Catholic reforms. John Winthrop, the Younger, was

G. Patterson, *Chemistry in 17th-Century New England*,
SpringerBriefs in History of Chemistry,
https://doi.org/10.1007/978-3-030-43261-4_2

well aware of the history of the English reformation. He also needed to navigate the treacherous waters of religious controversy. He learned his lessons well and was neither banished nor executed.

Another Puritan "hero" was John Hooper (1495–1555). He also chose to leave England to follow his faith in Switzerland. He was an admirer of Zwingli and had many contacts with Heinrich Bullinger and Jan Laski. He even married a Belgian woman. He eventually returned to England and was consecrated Bishop of Gloucester in 1551, even though he objected to the clerical vestments required of bishops. He took his job seriously and tried to "oversee" the clergy under his responsibility. Not surprisingly, most of them were ignorant and lazy. He believed that bishops should observe a vow of poverty! (definitely a sin worthy of death) When Mary became queen, Hooper was burned at the stake in 1555. John Winthrop, Jr., knew that there was a ground of commonality between all true Puritans, no matter where they lived. This was the commitment to the biblical evangelical faith and a deep distrust of clerical systems. While Cranmer desired to "maintain the unity of the faith," he chose to try and enforce a false uniformity in the name of the crown.

There was substantial religious strife in Scotland as well. At the start of the sixteenth century the official religion was Roman Catholic, but reformers such as George Wishart (1513–1546) promoted a more Reformed stance. His most famous disciple was John Knox (1513–1572). Wishart was charged with heresy in 1538 and fled to England. There he was "examined" by Archbishop Cranmer and forced to flee to Germany and Switzerland. In 1543 he returned to Scotland and carried out an itinerant ministry denouncing the Papacy. He was forced to stay on the move, one step ahead of the religious authorities. He was seized by Lord Bothwell and turned over to Cardinal Beaton, who had him hanged and burned in 1546. Wishart appears in *Foxe's Book of Martyrs*. He advised John Knox to keep a low profile, but when Beaton was murdered in revenge for killing Wishart, Knox was called to the Castle of St. Andrews to preach to the rebels. His reward was to be captured by the French and sentenced to the galleys. He was finally released in 1549, but had to flee to England. He was even installed as an Anglican preacher by Cranmer, now the Archbishop of Canterbury. But, religious intrigue forced Knox to resist preferment in the corrupt Anglican Church.

Under Queen Mary I, no Puritan was safe in England. Knox fled to Geneva and carried out a long dialogue with Calvin. He also interacted with Heinrich Bullinger. He published a pamphlet denouncing Mary Tudor and her Roman Catholic bishops. Many English exiles lived in Frankfurt, a free city, and they invited John Knox to be their minister. However, some of the English exiles wished to worship after the manner of the Anglican high church, including Edmund Grindal (1519–1583), the future Archbishop of Canterbury, and Richard Cox, one of the principal authors of the Book of Common Prayer. They stirred up trouble for Knox and he returned to Geneva. While he did visit his wife in Scotland, things could never be safe with both Mary and the bishops eager to arrest him. Only when Elizabeth ascended the throne of England did the exiles consider returning. Knox returned to Scotland to join the Scottish Reformation and the overthrow of the Roman Catholic Church. The death of Mary Guise, the Queen Regent, ended the rule of the Roman Catholic

Church in Scotland, and Knox and the Protestant Parliament proceeded to create the Protestant Kirk of Scotland. While John Knox was not technically a Puritan, his spirit of reformation and resistance to both ungodly civil and religious powers inspired the English Puritans. He was also thoroughly familiar with the major figures of the Reformation on the Continent. He was faithful to the end and inspired John Winthrop, the Younger, to keep his faith always in view, even as he navigated the dangerous political and religious waters of the seventeenth century.

While Queen Elizabeth I was a Protestant, she was also a monarch. She wished to be in control of all things in her realm. What she did achieve was a deep division between the high church Anglicans and the Puritans. Some of the Puritans defied the Queen, such as John Foxe (1516–1587). His monumental *Actes and Monuments* (1563) ("Foxe's Book of Martyrs") was known and read by all literate Puritan children (and all Puritans were expected to know how to read).

When it became clear that Queen Elizabeth and the high church prelates were determined to enforce "conformity," the public political party now known as the Puritans was born. They issued *An Admonition to the Parliament* (1572) that thoroughly inflamed the situation. Both sides still look back on this moment as a watershed: the Anglicans remember that Puritans cannot be trusted to "go along to get along," and the Puritans remember that the Anglicans were determined not to allow a complete Reformation to take place in England. This inflammatory document is still in print [4]! While it is hard for most people in the twenty-first century to comprehend the "vehemency" of this disagreement, no Puritan of the seventeenth century missed the point.

When James VI of Scotland became James I of England, the Puritans sought to influence the religious situation with a "Millenary Petition". After the coronation, James convened the Hampton Court Conference in 1604. While a few accommodations to the Puritans were made, soon thereafter the anti-Puritan Richard Bancroft was made Archbishop of Canterbury. One response of the "godly" was to emigrate to the Netherlands and the Pilgrims did just that [5]. They separated from the evil and corrupt Church of England. Another approach was to quietly extend the reach of the Puritan ministers within the Anglican communion. This was facilitated by the Puritan nobles who actually funded the local parishes. John Winthrop, the Younger, learned the value of being cooperative in official settings and quietly conducting his real Protestant reformation at home.

2.1   The Quiet Reformation

While the Anglican bishops and the Puritan clergy continued to battle, there was another stream that flowed within Puritanism. The Sovereignty of God was a central tenet of Puritan theology. While men pretended to determine the future, God had his own plans. When He was ready, God would act to bring forth a millennial future that was truly reformed. The story of *The Great Instauration* (1975) is brilliantly told by

Charles Webster [6]. This is the worldview that drove John Winthrop, the Younger, to invest his life in the pursuit of a millennial kingdom on Earth.

One of the key documents in this literary world was Sir Walter Ralegh's *The History of the World* (1614) [7]. It presented the story of mankind as a struggle between God and Antichrist with a triumphal victory of the "Kingdom of God". While the English church had chosen not to fully reform, God would choose a people who would be worthy of the Kingdom.

Another major literary figure in this story is Francis Bacon (1561–1626). Like Ralegh, Bacon wrote his best Puritan works while under arrest. Understanding the full Baconian program has been greatly assisted by Stephen Gaukroger's *Francis Bacon and the Transformation of Early-modern Philosophy* (2001) [8]. Francis Bacon was well-born and well-educated. He was an avid reader and a deep thinker. In addition, he was in personal contact with most of the best minds in England. He was a lawyer, and devoted much of his adult years to advising Queen Elizabeth or King James on legal matters. He was also a brilliant writer and was often the author of entertainments at the royal court. But, he wished to achieve true reform of learning so that England would prosper. How could he escape both the errors of the past and the treacherous situations in his present?

Francis Bacon believed that humans needed to take a new and unfettered look at all of reality. Most people grow up to have "opinions", but they do not know where they came from or how to think about them. Bacon believed that, since God was the author of reality, it was possible to obtain a clear view of all aspects of human life. He also believed that "a prudent question was half of wisdom". He may have been a little too optimistic, but his notion that by harnessing the full power of communities of committed scholars that reliable knowledge could be discovered is still the stance that causes scientists like me to get up in the morning. John Winthrop, Jr., understood both the need for spiritual guidance and the constraint of public knowledge judged by groups of active researchers. He tried to both join and create such groups. He was a founding member of the Royal Society, the only American to be so honored.

Naïve presentations of the Baconian paradigm have done great harm. It is a very sophisticated and complicated procedure, and it takes all of human history to achieve its goals. Once a true phenomenon has been discovered and verified, it remains true! Nevertheless, bad ideas from antiquity join the bad ideas of the present to obfuscate reality. Bacon taught that it is never safe to naively assume that anything is true unless it has been exhaustively verified. So, the first task is to assess prior understandings to identify those things that remain true. This is a very large task, and often receives no respect, but without clearing out the false ideas, it is impossible to see the correct ones. Some great scientists were especially good at detecting false ideas. Their first sense was often obtained intuitively, from a good knowledge of true things, but they did not rest until they had conclusively debunked the fallacious notions. The next step was to consider what else might be true, based on what was known to be correct. This step requires both good imagination, and rigorous attention to detail. If some notion is "assumed to be true," it is often easy to accept experimental results that might (at least on Thursdays) be accepted as valid. The notion that Bacon promoted "blind" exploration is fallacious. Experimental work is expensive and far too hard to waste time on random exploration. If the searcher has not been trained to "see reality," she usually just passes by the correct observation. This training is usually obtained as an apprentice with a skilled and experienced adept. John Winthrop, Jr., sought out the best chemists of his age, both in written form and in person. He carried out actual experiments with groups of alchemists. Eventually, he was the most respected adept in America.

Puritans also believed that every member of the Commonwealth was required for the success of the enterprise. The Massachusetts Bay Company was both a commercial venture and a religious community. John Winthrop, Jr., was fully committed to this effort and gave his life to it. While he was one of the most remarkable individuals of the seventeenth century, he never forgot the circles of other humans in which he traveled.

References

1. Miller P (1939) The New England mind: the seventeenth century. Harvard University Press, London
2. Westfall RS (1958) Science and religion in seventeenth-century England. Yale University Press, London
3. Lloyd-Jones DM (1987) The puritans: their origins and successors. Banner of Truth Trust, Edinburgh
4. Frere WH, Douglas CE (1907) Puritan manifestoes: a study of the origins of the puritan revolt with a reprint of the admonition to the parliament and kindred documents, 1572. Society for Promoting Christian Knowledge, London
5. Lamont WM (1969) Godly rule: politics and religion 1603–1660. St. Martin's Press, London
6. Webster C (1975) The great instauration: science, medicine and reform 1626–1660. Duckworth Books, London
7. Ralegh W (1829) The history of the world. Oxford University Press, Oxford
8. Gaukroger S (2001) Francis Bacon and the transformation of early-modern philosophy. Cambridge University Press, Cambridge

Chapter 3
The Errand into the Wilderness

Abstract A brief history of the founding of New England is presented. The key players are introduced in more detail, and the arrival of John Winthrop, Jr., is placed context. The role of alchemy in 17th century New England is explained and placed in the context of the Puritan Commonwealth.

Early attempts to colonize North America were fraught with peril. There were many dangerous aspects of Virginia, as documented by John Smith [1]. But, the biggest challenge associated with the Jamestown colony was the colonists themselves. They were truly unsuited to found a civil colony in a new world. Most of them had miserably failed in the Old World, and were not motivated to invest themselves in building a new one. Many of them were much more interested in the chemical gold, than in laboring after anything else.

John Smith continued to explore the coast of North America and became convinced that New England would be a better place to try and found a commercially viable colony. What was needed were people worthy of such a venture. The Puritans in England were wealthy, educated and adventurous. But, they were not popular in the court of James I or later Charles I. One solution to this problem was to immigrate to New England, where a commonwealth of elect saints could work for the glory of God and the benefits of one another. John Winthrop, Sr., had both the vision for such a project, and the monetary connections to lead a commercially viable entity: The Massachusetts Bay Company [2].

The history of New England from 1620–1649 is both more complicated and more interesting than the Middle School myth most of us were taught. The "Pilgrims" of Plymouth established a colony starting in 1620, but they had a longer history in the Separatist movement in England and as expatriates in Holland [3]. They were part of the general Puritan project, but they made themselves *persona non grata* in "Merrie Olde England." In order to land in New England, they needed permission from the people who had paid the King for the right to extort money from later immigrants. Fortunately, they had rich Puritan allies, and landed legally at Plymouth (Fig. 3.1).

Crown affairs in that part of the world were placed in the hands of the Council for New England in 1620. By the time the Pilgrims landed, they needed to apply to this new group for permission to be there (at additional cost). Patronage never ends, but

G. Patterson, *Chemistry in 17th-Century New England*,
SpringerBriefs in History of Chemistry,
https://doi.org/10.1007/978-3-030-43261-4_3

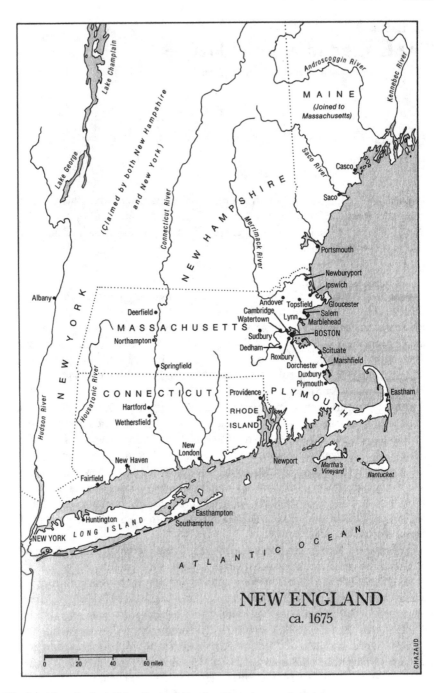

Fig. 3.1 Map of colonial New England (Stephen Innes, by permission)

each new administration re-establishes its network of privilege. The Pilgrims, led by William Bradford, needed to settle things with their proxy, John Pierce. This resulted in years of legal wrangling. "Free" land almost always comes with legal problems, and no one asked the resident peoples for their permission. John Winthrop, Jr., was scrupulous in his dealings with the Native Americans, and lawfully purchased rights to lands he occupied.

The Pilgrims had two major problems: (1) they knew nothing about how to survive in this forbidding and hostile land, and (2) in spite of pretentions to unity, they were a mixed multitude with multiple allegiances. Many of the original settlers died. Of those that survived, attempts to establish a working community led to conflicts over religious freedom and civil authority. While William Bradford's *History of the Plymouth Settlement* paints a "religiously correct picture of the pious Pilgrims," the dissidents published four major concerns: (1) There was in fact considerable diversity about religious questions, (2) rigorous "Sabbath Laws" interfered with family life, (3) the Sacraments were not offered in Plymouth (a real problem for more traditional Anglo-Puritan worshippers), and (4) no education for children in neither traditional grammar school subjects nor the catechism. The moneyed Pilgrims still in Europe sent a minister, Lyford, to Plymouth to administer the Sacraments according to the *Book of Common Prayer*. The Separatist members of the Pilgrim congregation had rejected all "Romish" practices, including the *Book of Common Prayer*. Eventually, several prominent members of the Plymouth group joined the Massachusetts Bay Company. Religion was parsed "very fine" in New England and no one could count on being on the "right" side in the future.

One of the consequences of this disagreement was the establishment of a colony on Cape Ann (Salem, Fig. 3.1) by a group of non-Separatists in 1623. This group was eventually led by the Reverend John White (1575–1648), even though he remained in England at his "conforming" parish. He arranged for groups of conforming Puritans to immigrate to Cape Ann. By the time the Plymouth adventurers led by Miles Standish (1584–1656) visited Cape Ann in 1624, there was already an established village there. This area became the Colony of Salem. White eventually joined the Massachusetts Bay Company, but remained in England as a friend and agent.

While commercial success eluded the early Salem community, continued enthusiasm for New England persisted in England. The next group of adventurers bought out the remaining rights of the Salem group and formed the New England Company for a Plantation in Massachusetts Bay. The aristocratic head of this group was Sir Richard Saltonstall (1586–1661, Fig. 3.2).

Saltonstall eventually led a group to New England himself and founded a plantation at Watertown in 1629. However, the weather did not suit him and he returned to England in 1631. Nevertheless, the name of Saltonstall persists in New England to this day.

The original New England Company was merged with the Massachusetts Bay Company in 1629. Puritans now began to immigrate in great numbers to New England (10,000 people in the period from 1629–1640). Some came from the West country areas like Dorchester, but John Winthrop and his people were from the Eastern counties like Suffolk, Essex and Lincolnshire. Not all Puritans were "the same."

They all opposed the clericalism of Archbishop Laud but some were loyal to the Church of England and some separated officially from this "corrupt body." The history of New England throughout the seventeenth century constantly reflects this religious tension. It is a miracle that John Winthrop, Jr., managed to survive the religious wars and the Indian wars in his lifetime.

Another important personage from this era was the 4th Earl of Lincoln, Theophilus Clinton (1600–1667). He was very interested in the Puritan cause. While he ended up in the Tower of London, most of his family immigrated to New England. One of his relatives was Thomas Dudley (1576–1653), often Governor of Massachusetts. Dudley was chosen as the deputy governor for the 1630 voyage on the *Arbella*, with John Winthrop, Sr., as governor. Their relationship over the next 20 years was sometimes contentious, but overall they remained good friends and served as useful debating partners. Exactly how to do what the Puritans wanted to do in New England was not known, and all good ideas needed to be debated. Dudley was a skilled lawyer and a hardened soldier. It took sturdy souls to fashion the Commonwealth in the Wilderness.

Fig. 3.3 Roger Williams
(1603–1683) (New England
Historical Society, by
permission)

Any short history of the foundation of the Massachusetts Bay Colony would be incomplete without a mention of the minister John Cotton (1585–1652). Cotton's academic career at Cambridge was brilliant. He entered Trinity College at age 13 and received his B.A. degree in 1603. He was a Fellow of Emmanuel College until 1622, and obtained M.A. and B.D. (Bachelor of Divinity) degrees. He was appointed to St. Bodolph's Church in Boston, Lincolnshire. One of the most significant events during his early years of ministry was meeting Anne Hutchinson (1591–1643). John Cotton was a Puritan, a stance gained from years at Emmanuel College, Cambridge. He preached the farewell sermon at the launching of the *Arbella* in 1630, but decided to remain in England. His growing conviction that the Anglican Church needed serious reformation came to the attention of the Canterbury authorities, and Cotton went into hiding. John Winthrop, Sr., again encouraged him to immigrate to New England and arranged to smuggle him out of England in 1633. He was immediately installed at the First Church of Boston, Massachusetts. Anne Hutchinson and her husband followed soon afterwards.

The subject of religious tolerance in New England is filled with irony. Roger Williams (1603–1683, Fig. 3.3) became the minister in Salem in 1634. He had traveled to New England on the same ship as John Winthrop, Jr. He argued that the Church of England was corrupt (hardly an issue in Puritan New England) and that good Christians should withdraw from it. For his troubles, John Cotton and the rulers of the Massachusetts Bay Colony threatened to send him back to England to face charges. Williams was "tolerant," but not of Anglicans. Cotton had been driven to America by intolerant Anglicans, and joined in the persecution of Roger Williams. John Winthrop, Jr., had a long and profitable relationship with Roger Williams.

John Cotton also survived what should have been his "finest hour," but instead led to the banishment of his best friends and colleagues: Anne Hutchinson and her brother-in-law, the Reverend John Wheelwright (1592–1679). They were all Puritans! What distinguished them was their embrace of the Grace of God. John Cotton was infamous for preaching the grace of God in Boston, Lincolnshire. How did he escape censure for preaching it in Boston, Massachusetts? Why would he have been censured: For suggesting that most of the clergy in New England were not in possession of "eternal life." Anne Hutchinson, having been taught well by John Cotton, promoted the notion that humans needed to accept the "free grace of God" in order to be admitted to the state of the elect. Standard Puritan (Anglican) theology stressed that the "elect" were chosen based on their life of devotion to the outward signs of spiritual life. Cotton characterized this approach as a "religion of works." While these arcane religious issues seem distant from the present, they were raging in New England. Calling the majority of the Puritan clergy unregenerate tended to produce rage. And they responded by bringing the Reverend Wheelwright before the Court of Assistants. He was banished from Massachusetts and went on to help found New Hampshire. With her brother-in-law in exile, and her mentor covering his behind, the Court next banished Anne Hutchinson in the dead of winter. Another of her political allies, Sir Henry Vane (1613–1662, Fig. 3.4), the Governor of Massachusetts in 1636, returned to England in 1637 when he was defeated by John Winthrop, Sr. He played a major part in the English government under Cromwell, but was beheaded in 1662 when Charles II was made King. Vane was a very tolerant man, and brought a sense of justice to the Parliamentarian era. He supported Roger Williams in the founding of Rhode Island. When Anne Hutchinson was exiled in 1638, Vane interceded for her with Williams. While Vane spent only a few years in New England, he was a force for good and was respected by John Winthrop, Jr.

But, why is the history of the Puritans in New England of any interest to chemists?! Chemistry in seventeenth century New England was alchemy. It was based on standard alchemical books by people like Paracelsus [4–6] and Libavius [7]. It was practiced according to the training he received in London, Amsterdam and Frankfurt. While there were standard books discussing the theory and practice of alchemy, the secret to understanding Puritan alchemy was the notion that success would come to those who labored at the altar of Vulcan (the furnace) and were blessed by God for their piety. Only the "godly" would achieve mastery and become adepts [8]. This mixed philosophy also influenced their understanding of medicine and witchcraft. They believed that only God actually heals. Nevertheless, iatrochemists were expected to employ chemicals in their practice, in addition to prayer and fasting. They believed that there were actual witches among the people, but it took great wisdom to identify who they were. John Winthrop, Jr., used his reputation as a genuine adept to exonerate many poor souls accused of witchcraft. In addition to adventitiously effective medicines, chemistry saved people from being burned at the

Fig. 3.4 Sir Henry Vane
(1613–1662) (Portrait by Sir
Peter Lely, public domain)

stake or being drowned. While the occasional alchemist in New England ran the risk of being accused of "black magic," Winthrop was able to protect this activity by his careful behavior and eloquent arguments. The tradition of admiration for chemists in America has not persisted into the twenty-first century!

References

1. Smith J (1907) The general historie of Virginia, New England and the Summer Isles. Macmillan, New York
2. Rose-Troup F (1930) The Massachusetts Bay company and its predecessors. Clearfield, New York
3. Bradford W (1920) Bradford's History of the plymouth settlement, 1620–1650, rendered into modern English by Harold Paget. Paget H (ed) E.P. Dutton and Company, New York
4. Webster C (2008) Paracelsus: medicine, magic and mission at the end of time. Yale University Press, New Haven
5. Westfall RS (1958) Science and religion in seventeenth-century England. Yale University Press, New Haven
6. Debus AG (1977) The chemical philosophy: paracelsian science and medicine in the sixteenth and seventeenth centuries. Neal Watson Academic Publications, New York
7. Moran BT (2007) Andreas Libavius and the transformation of alchemy: separating chemical cultures with polemical fire. Science History Publications, Sagamore Beach
8. Young JT (1998) Faith, medical alchemy and natural philosophy: Johann Moriaen, reformed intelligencer, and the Hartlib Circle. Ashgate, Aldershot

Chapter 4
John Winthrop, Jr.: The Making of an Adept

Abstract The life of John Winthrop, the Younger, is presented in the context of his progress as an alchemical adept. He collected the finest library in New England in the 17th century and alchemical books were the dominant genre. He was in communication with most of the working alchemists of his time and especially admired John Dee. He traveled throughout the European world and made friends wherever he went. By the time he immigrated to America he was a recognized adept and was able to apply the principles of alchemy to his daily activities. He also established himself as an Assistant of the Massachusetts Bay Company.

John Winthrop, Jr. was born on February 12, 1605 at Groton, Suffolk. Since he was the son of a landed squire, he was educated to attend college at the celebrated Free Grammar School founded by Edward VI at Bury St. Edmunds, Suffolk [1]. The Puritan aristocracy existed alongside the more worldly courtiers and there were many men of substance. When he was 16 years of age, John attended Trinity College, Dublin at a time when great scholars like Sir William Temple (1555–1627), Provost, and Archbishop James Ussher (1581–1656) were associated with Trinity. He was under the care of his uncle, Emmanuel Downing (1585–1658), during this period. He studied hard for a year, but with the impending return of his uncle to London, John Winthrop, Jr., returned home to Groton, much to the displeasure of his father. He did make a good impression on his tutor, Joshua Hoyle, and they remained friends.

With the help of his uncle, John went to London and became a barrister in the Inner Temple. While the law itself held no great interest for John Winthrop, Jr., he did meet many people who would matter to his future while in London. For example, he made the acquaintance of Edward Howes, and a joint interest in alchemy led them to carry out many experiments. They remained lifelong friends and correspondents. Howes also helped Winthrop obtain alchemical books and shipped many to New England. Another prominent source for both books and chemicals was Francis Kirby.

What sorts of chemistry books were available in 1631? The classic bibliography of chemistry by Henry Carrington Bolton [2] lists volumes such as:

1. *Of the Vanitie and Uncertainty of Arts and Sciences*, Agrippa von Nettesheim (1569)

© The Author(s), under exclusive license to Springer Nature Switzerland AG 2020
G. Patterson, *Chemistry in 17th-Century New England*,
SpringerBriefs in History of Chemistry,
https://doi.org/10.1007/978-3-030-43261-4_4

2. *De Mineralibus*, Albertus Magnus (1569)
3. *Theatrum Chemicum*, (1613–1622)
4. *Chymische Hochzeit Christiani Rosencreutz, Anno 1459*, Johann Valentin Andrea (1616)
5. *The Mirror of Alchemy*, Roger Bacon (1597)
6. *Tractatus de veritate alchemiae*, Adam von Bodenstein (1560)
7. *Basilica Chemica*, Oswald Croll (1609)
8. *Apologia chrysopoeiae contra Thomam Erastum*, Gaston de Clavius (1598)
9. *Extractum chymicarum*, Johann Conrad Gerhard (1616)
10. *Dyas chymica Tripartita*, Johann Grasshof (1625)
11. *Libelli aliquot chemici*, Raymond Lull (1515)
12. *Atalanta fugiens*, Michael Maier (1618)
13. *Apologia chemical adversus invectivas Andrae Libavii calumnias*, Joseph Michelius (1597)
14. *Museum Hermeticum* (1625)
15. *Magia naturalis*, Johannes Baptista Porta (1562)
16. *The Compound of Alchemy*, George Ripley (1591)
17. *Lexicon alchemiae*, Martin Ruland (1612)
18. *Tripus Chemicus Sendivogius* (1626).

The nature of the complete library of alchemical books collected by John Winthrop, Jr. has been explored by Wilkinson [3]. A list of 275 such books was compiled, based on actual books in the libraries of Yale University, the New York Academy of Medicine, the New York Society Library and several smaller collections. Many of the books and manuscripts are not listed in Bolton, but may be found in other bibliographies such as the *Bibliotheca Chemica* compiled by John Ferguson (1954) [4]. A few of the earliest ones are:

1. *A Revelation of the Secret Spirit*, Giovanni Baptista Agnello (1596)
2. *Tractatus Nobilis Primus*, Peter Amelung (1608)
3. *Chymische Hochzeit Christiani Rosencreutz, Anno 1459*, Johann Valentin Andrea (1616)
4. *Medicinae Chymicae*, Francis Anthony (1610)
5. *Tractatus Chemicus*, Arnaldus de Villa Nova (1504)
6. *De Occulta Philosophia*, Basilius Valentinus (1603)
7. *Tyrocinium Chymicum*, Jean Beguin (1610)
8. *Tractatus de veritate alchemiae*, Adam von Bodenstein (1560)
9. *Introductio*, Petrus Bonus (1572)
10. *Basilica Chemica*, Oswald Croll (1609)
11. *Dispensatorium Chymicum*, (1626)
12. *Congeries Paracelsicae de Transmutationibus Metallorum*, Gerhard Dorn (1531)
13. *Clavis Philosophiae et Alchymiae Fluddanae*, Robert Fludd (1633)
14. *Harmoniae Imperscrutabilis Chymico-Philosophicae*, Johann Grasshoff (1625)
15. *Alchemiae*, Gratarolo Guglielmo (1572)
16. *Alchimia*, Petrus Kertzenmacher (1570)

17. *De Igne Magorum Philosophorum,* Heinrich Khunrath (1608)
18. *Alchemia,* Andreas Libavius (1597)
19. *Libelli aliquot chemici,* Raymond Lull (1515)
20. *Arcana Arcanissima,* Michael Maier (1614)
21. *Meteororum,* Paracelsus (1566)
22. *Magia naturalis,* Johannes Baptista Porta (1562)
23. *The Compound of Alchemy,* George Ripley (1591)
24. *Lexicon alchemiae,* Martin Ruland (1612)
25. *Anatomia Antimonii,* Angelo Sala (1617)
26. *Novum Lumen Chymicum,* Michael Sendivogius (1628)
27. *Idea Medicinae Philosophicae,* Petrus Severinus (1571)
28. *Antimonii Mysteria Gemina,* Alexander von Suchten (1604)
29. *Metallurgia,* Joachim Tancke (1605)
30. *Theatrum Chemicum,* (1613–1622)
31. *Tractatus de Secretissimo Antiquorum Philosophorum Arcano,* (1611).

John Winthrop, Jr., was a truly serious collector of alchemical books. He did much more than just pile them on his shelves. He devoured them. His thirst for alchemical knowledge was great and he spent as much time as he could in pursuing alchemy, both in print and in the laboratory.

The practice of alchemy in the early seventeenth century was quite varied. There were many communities that impacted this activity. One group was the mining and metallurgy guild. Clear descriptions of this industrial activity are contained in the classic book by Georgius Agricola (1494–1555) *De Re Metallica* (1556) (Fig. 4.1).

Another large group was the pharmacists and the Paracelsian physicians. A good example of this literature was the work of Oswald Croll (1560–1609) *Basilica Chimica* (1609) (Fig. 4.2).

The full *materia medica* was known to Winthrop. And, finally, the *arcana* of the alchemists was well represented in his library. For example, books by Gerhard Dorn (sixteenth century) *Clavis Totius Philosophiae Chymisticae* (1583) (Fig. 4.3).

As will become clear, John Winthrop coupled his reading and his laboratory work with frequent correspondence with members of the worldwide alchemical fraternity. He was also able to visit with most of the known alchemists in his lifetime. Although he was never able to meet John Dee (1527–1608) in person, he did receive a copy of one of his books directly from Dee's son. His favorite book by Dee was the *Monas Hierogpyphica* (1564) (Fig. 4.4).

Winthrop adopted the symbol shown in Fig. 4.4 as his trademark and it appears on many of his trunks and papers (Fig. 4.5).

John Dee (Fig. 4.6) was one of the greatest chemical bibliophiles of all time. John Winthrop was definitely one of his disciples. Like Dee, two principles guided him: (1) chemical truth is found in the laboratory, not just in books, and (2) God chooses to communicate arcana to a few men who are prepared to receive this knowledge. Since it is a gift, it must be valued and protected.

John Dee was a student at the same King Edward VI Grammar school, and John Winthrop, the Younger, would have learned of this almost legendary figure there.

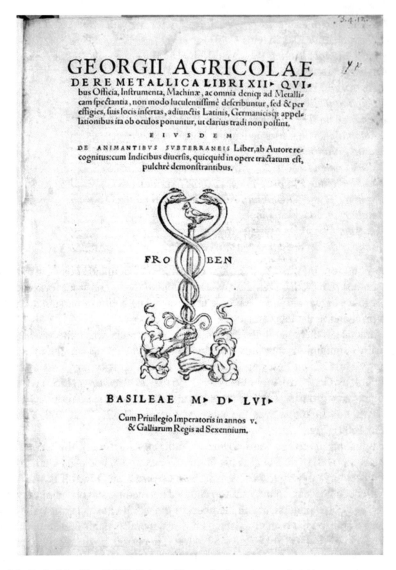

Fig. 4.1 *De Re Metallica* (1556) (Science History Institute, by permission)

Dee graduated from St. John's College, Cambridge and went on to become one of the original Fellows of Trinity College, Cambridge at its founding by Henry VIII in 1546. He was a serious scholar of Euclid and produced a critical edition of *Elements* in 1570. He traveled extensively in Europe and presented a copy of his *Monas Heiroglyphica* to the Holy Roman Emperor Maximillian II. He bought manuscripts and books wherever he went.

Fig. 4.2 *Basilica Chymica* (1609) (Science History Institute, by permission)

4.1   Wanderjahre

The life of a lawyer did not suit John Winthrop, the Younger's adventurous nature and his uncle helped him obtain a position as secretary to Captain Thomas Best (1570–1639) of the ship of war *Due Repulse*. This expedition was an attempt to liberate

Fig. 4.3 *Clavis totius philosophiae chymicae* (1583) (Science History Institute, by permission)

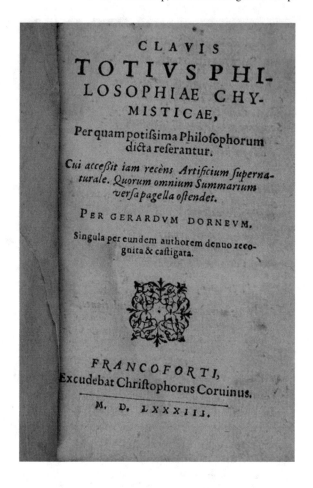

the French Huguenots who were under siege in New Rochelle, France. While the naval failure of this voyage meant that he needed to move on, he had tasted adventure and purposed to join the Puritan immigration to New England in 1629. He also met Cornelius Drebbel (1572–1633) and his son-in-law Abraham Kuffler (1598–1657). They were very active chemists and played a central role in the chemical life of both England and the Netherlands. They were engaged by the Duke of Buckingham as munitions experts.

Cornelius Drebbel was a good example of the type of alchemist practicing in the Protestant world of the early seventeenth century. He was Dutch, but traveled to wherever a source of funds could be found. He moved to England in 1604 at the invitation of James I and was attached to the court of Henry, Prince of Wales. He was a polymath and inventor and entertained the royal court with remarkable machines involving mechanics, hydraulics, alchemy and optics. He was also invited by Emperor Rudolph II to demonstrate his alchemy and his inventions. He founded a dye company with his sons-in-law, Abraham and Johann Sibertus Kuffler, that

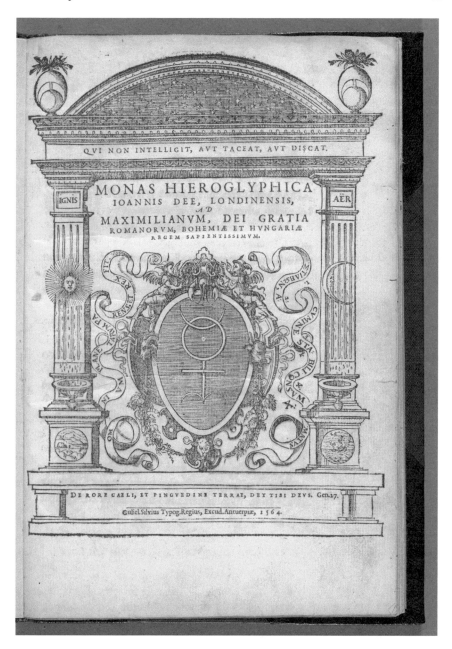

Fig. 4.4 *The monas heiroglyphica* by John Dee (Huntington Library, Pasadena, CA by permission)

FIGURE 1. *Monas Hieroglyphica*, by John Dee. Detail of page 12. Antwerp, 1564.
Courtesy of the Massachusetts Historical Society

Fig. 4.5 The Monas Heiroglyphica of John Dee (Science History Institute, by permission)

Fig. 4.6 John Dee, advisor to Queen Elizabeth I, and inspiration for John Winthrop, Jr. (public domain)

produced a famous carmine color based on stannous chloride and cochineal. John Winthrop, Jr., carried on an extensive correspondence with Abraham Kuffler, and they interacted in person on his trips to England.

John Withthrop's father cautioned him about being too hasty in his decision to emigrate, so John Jr., followed his *wanderlust* as a paying passenger on a merchant ship plying the Mediterranean. He made many stops at places like Tuscany, Pisa, Florence, Venice and Constantinople. He sought out alchemists wherever he went and initiated correspondences that lasted his entire life. He became familiar with many cultures and learned to get along with many different kinds of people. While he was a very devout Puritan, he was very tolerant of the people who shared his passions for chemistry and politics. As always, he made friends with the people he met. One of them was Sir Peter Wyche (1593–1643), the English envoy to the Ottoman "Porte [1]." Another was Sir Isaac Wake (1581–1632), the English ambassador to Savoy and Venice. But, the most important for the present story was Jacobus Golius (1596–1667), a Dutch scholar from Leyden who had obtained many alchemical and mathematical manuscripts in the Middle East.

4.2 The Massachusetts Bay Company

John Winthrop Sr., did join the Massachusetts Bay Company and became their Governor in 1629. Upon hearing that his father was committed to this enterprise, John Jr., wrote to him [5].

"I believe confidently that the whole disposition thereof is of the Lord, who disposeth all alterations, by his blessed will, to his own glory and the good of his; and therefore do assure myself that all things shall work together for the best therein."

In the summer of 1629, the Massachusetts Bay Company was seriously considering "immigrating to America" and by August 26, both Emmanuel Downing and John Winthrop, Sr., had joined the company and committed to the adventure. By October 20 John Winthrop Sr., had been elected Governor, instead of a worthy like Sir Richard Saltonstall. Plans were now put in full motion to prepare for the voyage. Since Senior needed to be in London, as Governor of the Massachusetts Bay Company, Junior was placed in charge of the estate in Groton. He also carried out many necessary tasks, such as recruiting skilled craftsmen. Unlike the Pilgrim voyage to America, the Puritans made sure to include all the trades necessary to build a commonwealth in the wilderness, including good brewers. John Winthrop, Jr., also made inquiries about building forts and sketched some of the best in England. His ability to envision new things and plan their execution was demonstrated many times in the future.

The pilgrimage to New England was set for April of 1630. John Winthrop Sr., took along two of his younger sons, Stephen and Adam. John Winthrop, Jr., needed to remain in England to deal with the family business and to arrange for the sale of the property in Suffolk. During this period he also married and prepared to join his father and their co-religionists in Massachusetts. This was a major undertaking and required the outfitting of another ship, the *Lion*. In all these affairs, as in so many before, his uncle, Emmanuel Downing, was of great assistance and was now an officer of the Company. John Winthrop, Jr., continued to meet and impress the Puritan aristocracy. His father was so pleased that he wrote "Among many of the sweet mercies of my God towards me in this strange land, … he hath given me a loving and dutiful son".

In 1631, John Winthrop Jr., sailed for New England with his wife, his step-mother and her two children. In addition to many supplies for the Puritans already in New England, the passengers brought along their household goods. John Winthrop, Jr., shipped a barrel of "chimical books." Amidst his frenetic activities in support of the Massachusetts Bay Company, his many trips to London allowed him to interact with his alchemical friend, Edward Howes, and to meet other leading Puritans. The *Lion* was greeted with great joy and martial celebration upon its arrival in Boston. It also contained an eager young minister named Roger Williams and a scholarly young Essex clergyman named John Eliot (1604–1690). Eliot devoted years to the evangelization of the Native Americans and was celebrated in Morison's *Builders of the Bay Colony* [6].

After a short period of adjustment to the new land and its weather, John Winthrop Jr. found employment in arranging for the shipment and trading of goods for New England, a skill he learned during the planning of two Puritan voyages. He took advantage of his friend in London, Francis Kirby, to purchase both scientific books for himself, and necessary instruments for New England. He also carried on an extensive correspondence with his friends in England. This proved to be a real boon to the Puritans, since there were always adverse schemes being initiated by enemies among the court of Charles I. John Winthrop, the Younger, often discovered these schemes in time to fend them off, with the help of Emmanuel Downing and other highly placed Puritan peers. His father soon recognized his political skills and had him elected as one of the Assistants of the Massachusetts Bay Company. Rather than a case of nepotism, John Winthrop, Jr. proved to be one of the most able politicians in the history of New England. He helped to develop a dissembling policy of praising the King and the English church in London, but proceeding differently at home. Historically, this has been the essence of "diplomacy".

References

1. Black RC (1966) The younger John Winthrop. Columbia University Press, New York
2. Bolton HC (2005) A select bibliography of chemistry, 1492–1892. Martino Publishing, Mansfield Center
3. Wilkinson RS (1963) The alchemical library of John Winthrop, Jr. (1606–1676) and his descendants in Colonial America, Parts I-III. Ambix XI:33–51
4. Ferguson J (1954) Bibliotheca chemica. Derek Verschoyle Academic and Bibliographical Publications, London
5. Waters TF, Winthrop RC (1899) A sketch of the life of John Winthrop, the younger, founder of Ipswich, Massachusetts, in 1633. Ipswich Historical Society
6. Morison SE (1930) Builders of the Bay Colony. Cambridge University Press, Cambridge

Chapter 5
Ipswich: Founding a Town in New England

Abstract One of the most important things John Winthrop, Jr., did for New England was to found new towns. His first venture was a fishing village near Salem on Cape Ann, Agawam (later Ipswich). The land needed to be cleared, surveyed, parceled and houses built. Winthrop did his part, both in manual labor and in detailed planning and in political organization. His next big task was to recruit more Englishmen to immigrate to New England. He traveled end back to England and met many Puritan leaders. A successful recruiting trip included many figures that were to play a decisive role in Massachusetts and Connecticut history. John Winthrop, Jr., also made great progress in his desire to become an alchemical physician. Upon his return he founded the fort at Saybrook, Connecticut. He then returned to Ipswich and founded a saltworks for New England. During this time period he met a highly educated and wealthy alchemical physician named Robert Child during his tour of the Americas in search of opportunities. They remained good friends and he played an important part in the history of New England.

After John Winthrop, Jr. had settled in and proved his value, he was chosen to start a new village at the site of the Indian village of Agawam in 1633, afterwards called Ipswich [1] (Fig. 5.1). The motivation for settling the land surrounding Cape Ann was the aggressive presence of the French. Winthrop had learned that they were planning to annex as much coastland as they could to add to their land of Acadia, in what is now Canada. This site, 30 miles northwest of Boston, located at the mouth of the estuary of the river now called the Ipswich, was judged to be "the best place in the land for tillage and cattle."

Rather than merely squat on the land, John Winthrop, Jr. negotiated the purchase of the property with Musconominet, the Sagamore of Agawam. After the village was settled and the houses occupied, Agawam was rechristened Ipswich. Ipswich, England was in Suffolk County and was well-known to its new Colonial inhabitants. Soon thereafter, in 1634, Nathaniel Ward (1578–1652) was installed as the minister. He was a Puritan minister deprived of his pulpit by Archbishop Laud. England's loss was Massachusetts' gain. Ward had a background as a soldier, lawyer and clergyman, and served Massachusetts as a wise political counselor. He is remembered today for his lively and humorous pamphlet "The Simple Cobbler of Aggawamm" (Fig. 5.2).

G. Patterson, *Chemistry in 17th-Century New England*,
SpringerBriefs in History of Chemistry,
https://doi.org/10.1007/978-3-030-43261-4_5

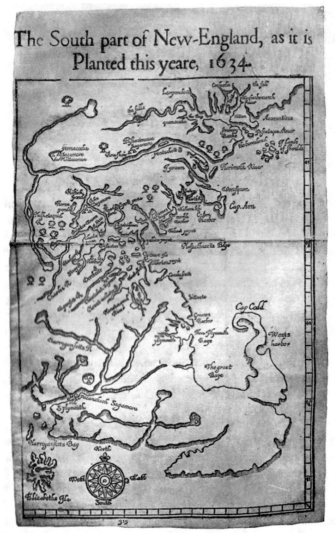

Fig. 5.1 Map of New England in 1634 (scanned from personal copy of *Builders of the Bay Colony*)

It took more than a year to clear the land and build houses so that the women and children could join the men. John Winthrop, Jr. endured this hardship and gladly welcomed his wife the next year. But, soon after the birth of his daughter, both she and his wife died and were buried in Ipswich. As one of the founders of the town, John Jr. continued to play a role in its development, but within a year of his wife's death, he returned to England.

Fig. 5.2 Title page of
Simple Cobler of Aggawamm
(scanned from my personal
copy)

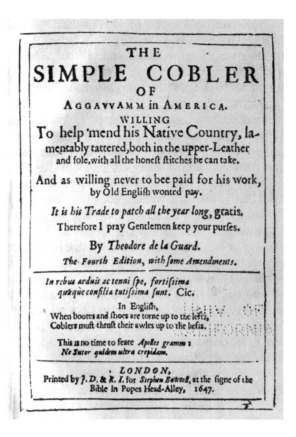

5.1 Recruiting Trip to England

The motivation for this trip is shrouded in mystery, but recent research has revealed that it was a recruiting trip to encourage more Puritans to emigrate to New England [2]. Another important political purpose of the trip was to defeat efforts by three groups of enemies to take over the growing and thriving enterprise: (1) The Council for New England existed to benefit the members of the Council, and Sir Ferdinando Gorges (1565–1647) had obtained a grant that included all of New England, (2) The Crown had received very little benefit from New England and was eager to insinuate itself into the Colony, and (3) The Church of England, under the leadership of William Laud (1573–1645), was eager to replace all nonconforming ministers with loyal High Church Anglicans and to deprive all Puritan ministers of their living.

Sir Ferdinando failed in his attempts to steal New England, but not without skillful maneuvers by Emmanuel Downing. Gorges was a decorated soldier and Governor of the Port at Plymouth, England. He was a distant relative of Walter Raleigh, and was part of the Plymouth Company that attempted a colonization of Maine in 1607!

He received a "land patent" for New England between the Merrimack and Kennebec rivers from King James I in 1622.

The current English King soon had other problems that were more pressing than a wilderness 3000 miles away. While Charles I (1600–1649) had signed the charter for the Massachusetts Bay Company, he was not favorably disposed to the Puritans in general. He had married a French Catholic (Henrietta Maria) and was a strong promotor of William Laud in his attempts to create a High Church Anglican monopoly. He was often too smart by half in his dealings and betrayed many of his courtiers. In addition to his troubles in England, there were Irish rebellions throughout the 1640s. And the Scots were available both to be bribed by Charles I and the Parliamentarians.

William Laud would soon be led to the block by Sir John Clotworthy (d. 1665), a loyal Puritan that John Winthrop, Jr. met on his travels in Northern Ireland. Clotworthy was a Puritan's Puritan. Although he was not an ordained minister, he was a flaming partisan. He is most infamous for two outrageous acts: (1) He destroyed a famous painting of the Crucifixion by Rubens that hung in Queen Henrietta Maria's private chapel, and (2) He harangued William Laud in the moments before his beheading. His family was well-known in Anglo-Irish politics and he served as MP (Member of Parliament) in the Irish Parliament for County Antrim. He also sat in the Short and Long Parliaments in England as the representative of Cornwall. He was a vehement hater of Thomas Wentworth, 1st Earl of Strafford (1593–1641), and hectored him to an impeachment and execution. Lord Strafford had been the Lord Deputy of Ireland under Charles I. Clotworthy participated in the trial of Archbishop Laud and promoted his execution. He fell out with the Parliamentarians and spent three years under arrest in Wallingford Castle. He was an Irish Presbyterian, but he helped Charles II regain the English throne and served the Crown in Ireland.

While John Winthrop, Jr. was in Ireland and England he was active in recruiting Puritans to come to New England. He was a natural salesman and he was truly committed to the American plantation. He continued to make friends with the Puritan aristocracy. He was commissioned by Lord Say and Sele and Lord Brooke, two close friends of his father, to found a plantation on the Connecticut River. One of the most important men he courted was Sir Matthew Boynton (1591–1647). He married a new wife, Elizabeth, the step-daughter of Hugh Peter (1598–1660), a noted Puritan divine, then in exile in Holland.

As in his first two adventures in planning a return voyage, he was busy buying goods for his father's family, his family, and for New England. Many of these were alchemical books and chemical apparatus. He also served as an agent of family and friends with regard to the disposition of property in England. While he was reputed to be weak in "accounting," he was definitely a man of account when it came to responsible agency.

Since John Winthrop, Jr., was definitely not a certified medical doctor with a degree from a school like the University of Edinburgh, it might be wondered how it was that he became the most famous physician in seventeenth century New England. Part of the answer was his interactions with Abraham and Johann Kuffler (1595–1677), two German doctors of medicine with a penchant for iatrochemistry. They were part of the extended circle of alchemists associated with figures like Cornelius Drebbel (1572–1633), Johann Glauber, Johann Morien and Benjamin Worsley. Winthrop read many of the known books on alchemical medicine and learned how to prepare the standard alchemical pharmaceuticals. He even concocted his own favorite remedy called "rubila." Morison [3] quotes Oliver Wendell Holmes on its composition, gleaned from a close perusal of the Winthrop papers in preparation for a Lowell Lecture: "Four grains of diaphoretic antimony; twenty grains of nitre; with a little salt of tin; plus a little red coloring to make it look more like a 'medicine.'"

A few of Winthrop's known medical books include:

1. Agricola, J. *Commentariorum, Notarum, Observationum & Animadversionum in Jophannis Poppii Chymische Medicin*, Leipzig (1638).
2. Anthony, F. *Medicinae Chymicae*, Cambridge (1610).
3. Beuther, D. *Der Medicin Doctoris Universal*, Frankfurt (1631).
4. Camerarius, J.R. *Sylloges Memorabilium Medicinae Et Mirabilium Naturae Arcanorum* (1624).
5. *Dispensatorium Chymicum* (1626).
6. Evans, J. *The Universal Medicine: Or the Vertues of the Antimonial Cup*, London (1634).
7. Fludd, Robert, *Medicina Catholica*, Frankfurt (1629).
8. Helmont, J.B. van *Oriatrike, Or, Physick Refined*, London (1662).
9. Hornung, J. *Cista Medica*, Nurnberg (1626).
10. Khunrath, C. *Medulla Destillationa Et Medica sextum aucta*, Hamburg (1639).
11. Liebaut, J. *Quatre Livres Des Secrets De Medicine* (from the library of John Dee) Rouen (1628).
12. Muller, P. *Miracula & Mysteria Chimica-Medica* (1614).
13. Mylius, J.D. *Opus Medico-Chymicum,* Frankfurt (1618).
14. Nollius, H. *Systema Medicinae Hermeticae Generale*, Frankfurt (1613).
15. Paracelsus, T. *Volumen Medicinae*, Strassburg (1575).
16. Penotus, B.G. *De vera preparation & usu Medicamentorum chemicorum*, in *Theatrum Chemicum* (1616).
17. Polemann, J. *Novum Lumen Medicum*, Amsterdam (1659) (Gift from Samuel Hartlib).
18. Renou, J. *Dispensatorium Medicum*, Geneva (1645).
19. Rhumel, J. *Medicina Spagyrica*, Frankfurt (1648).

One of his favorite books was *Of Natural and Supernatural Things* (1603) by Basil Valentine. His own copy was a gift from Abraham Kuffler and had belonged to Cornelius Drebbel (Fig. 5.3).

Since there were only a few medical schools in Europe, and only the privileged were likely to attend them in the seventeenth century, it required real initiative to become a successful physician. In addition to his medical library, John Winthrop, Jr., cultivated a large group of practicing iatrochemists so that he could adopt practices that were known to work. Isolated "empirics" tended to become little more than quacks, while John Winthrop, Jr., was admired as a physician because most of his patients lived! This can hardly be said of typical members of the Royal College of

Fig. 5.3 Title page of 1670 edition in the Othmer Library of the Science History Institute (by permission)

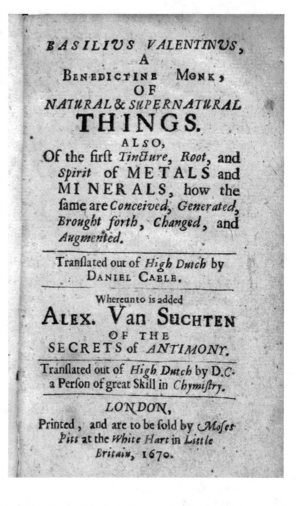

Physicians in the seventeenth century. One of the best-known of the Ipswich medical chemists was William Avory (also called Avery).

The return voyage to New England included a very large company of Puritans. Many of them traveled under assumed names since they were being sought by Archbishop Laud. The most notable were John Wilson (1588–1667), a famous Puritan preacher deprived by Laud who took the lead in prosecuting Anne Hutchinson; Hugh Peter (1598–1660) (his father-in-law) who presided at the beheading of Charles I and was treated to the same fate by Charles II; John Norton (1606–1663) who became a teacher in the church at Ipswich; John Jones (1593–1665) who became the minister in 1636 at Concord, MA and from 1644–1665 served at Fairfield, CT; and Thomas Shepherd (1605–1649) the pastor at Cambridge and a central figure in the founding of Harvard College. An even more notable passenger was Henry Vane, the Younger (Fig. 3.4), who was to become Governor of the Massachusetts Bay Company [4].

Vane played a prominent role in the Parliamentary government and ended up on Tower Hill! This extraordinary visit of John Winthrop, Jr. to England was a success, and it set the pattern for the rest of his life.

Soon after his return from England, John Winthrop, Jr. organized a party of men to build a fort at Saybrook, Connecticut, near the mouth of the estuary of the Connecticut River. He traveled to Saybrook the following year, 1636, but chose not to make his home there at that time. It was not clear how the town was going to thrive and the Puritan Lords and Gentlemen in England were occupied with a new venture: bringing down the King of England.

After considering his position in Massachusetts, he returned to Ipswich with his wife in 1637 and joined the rulers of the town. His decision to return to Ipswich was encouraged by a personal letter from Nathaniel Ward and an official letter from the leading men of the town to his father, Governor Winthrop. It expresses their desire that they be not "deprived of one whose presence is so grateful and useful to us." The letter is signed by 57 men, headed by Richard Saltonstall and Nathaniel Ward.

As part of his duties, he often needed to go to Boston for meetings of the Massachusetts Bay Company and the Court of Assistants. With all his responsibilities, one might wonder when he had time to pursue scientific matters, but he was often traveling and observed what he saw. He also made specific outings of the surrounding territory to look for minerals.

5.2 Saltworks

Importing salt from England or from the Caribbean was expensive. But, there was plenty of salt to be found in the oceans that bathed New England. One method used by the local Indians was to boil sea water in a large kettle until the salt crystallized out upon cooling. While this technique was suitable for small-scale production, it was not useful for producing large amounts of salt. There were no known salt mines in New England in Colonial times, and also no known salt springs; but, there was the Atlantic Ocean!

Large iron kettles were very expensive and had to be imported from England. New England was blessed with timber. While burning this timber in the small-scale salt making process was possible, a better use for the wood was to construct salt boxes (Fig. 5.4). Filtered sea water could then be evaporated by the sun on good summer days. It rains all summer long in Massachusetts. The solution is to provide covers for the salt boxes during the evening and when it rains. In 1638 John Winthrop, Jr. founded a salt making operation at the seashore near Ipswich at a place called Ryall-Side, now called Beverly. It was producing salt by the summer of 1639. This activity may not have inspired Robert Boyle to philosophize, but it is an example of the practical chemistry that motivated John Winthrop, Jr. He continued to be interested in salt making throughout his career in New England. The Chemist's Club (New York) continues to award the Winthrop-Sears Medal in honor of Colonial saltmaking.

Fig. 5.4 Saltmaking on Cape Cod (public domain)

In addition to the covered boxes, John Winthrop used windmills to pump the sea water into the saltboxes in some of his later saltworks in the Caribbean. Not only did this industry save New England, but it persists today in a thriving gourmet salt industry (Fig. 5.5).

Involvement in local industrial activities also allowed John Winthrop, Jr., to avoid being present for the religious trials of John Wheelwright and Anne Hutchinson. He detested religious controversy, and he did not wish to waste time on such matters. Instead, he produced salt and children. He also read the many books that he purchased and thought about alchemy. He needed to find a useful place in the life of New England besides just being a politician. He had a wife and family to support.

Another truly important activity was his correspondence with Europe. In addition to family news and alchemical recipes, he received intelligence about the progress in the war against Charles I. As the English Civil War took hold, emigration ceased, and the necessary people and capital required to sustain the growth of Massachusetts

Fig. 5.5 Modern salt from Massachusetts (by permission from Martha's Vineyard)

failed to materialize. Almost any colonial enterprise is a kind of pyramid scheme. Later joiners provide the capital to keep the game going. But, how then was New England going to survive: by becoming more independent. John Winthrop, Jr., had two things to offer to his compatriots in Massachusetts; (1) his political skills and wide knowledge of the Puritan aristocracy, and (2) his chemical knowledge and his ability to encourage men of substance to invest in chemical industry. These two factors led in 1641 to another voyage to England.

5.3 Robert Child (1613–1654)

One of the most interesting characters in the history of New England, and one who became a close friend of John Winthrop, Jr., was Robert Child. He was wealthy and highly educated. He obtained an M.A. from Cambridge in 1635 and was a Doctor of Medicine (1638) who attended both Leyden and Padua, two of the leading medical schools in Europe. In addition to his extensive travels in Europe, in 1639 he made a trip to New England to search for minerals. On this trip he met and developed a relationship with John Winthrop, Jr. They corresponded for the rest of his life. Child returned to England in 1640, but was determined to return to New England as an industrial entrepreneur.

While John Winthrop, Jr., was in England and Europe in 1642–1644, he interacted with Child. They both were members of the Hartlib Circle and planned to found an alchemical "plantation" in New England. But first, Child became one of the investors in the proposed ironworks (see Chap. 6). He also invested in the "black lead" (graphite) mine developed by Winthrop. With such good contacts, considerable capital, and more expertise in scientific matters than anyone in New England, one might suppose that he would be celebrated as a great American. But, his history took a strange turn.

Robert Child returned to New England in 1645. While he did invest in New England land and enterprise, he also expected to be taken seriously by the governors of Massachusetts. Robert Child was a true Puritan, but by 1645 there were many different sects of the "godly." Child was a Presbyterian, consistent with the Parliamentarian government, and expected the same rights and liberties as he would have had in England. Massachusetts Puritans were Congregationalists. Political franchise was contingent on religious conformity. When he joined the other "Remonstrants" in petitioning the General Court of Massachusetts for their rights to the franchise and to freedom of worship, they were heavily fined and jailed. Eventually, Child returned to England. He continued to correspond with John Winthrop, Jr., but plans to live at the plantation at New London, Connecticut were finished.

References

1. Waters, TF, Winthrop RC (1899) A sketch of the life of John Winthrop, the Younger, founder of Ipswich, Massachusetts, in 1633. Ipswich Historical Society
2. Black RC (1966) The Younger John Winthrop. Columbia University Press, New York
3. Morison SE (1930) Builders of the Bay Colony. Cambridge University Press, Cambridge
4. Willcock J (1913) Life of Sir Henry Vane the younger, statesman and mystic (1613–1662). Oswaldestre House, London

Chapter 6
Return to England (1641) and the New England Ironworks

Abstract One of the great needs of the Colonial enterprise was a local source of ironware. John Winthrop, Jr., traveled to England to raise the capital and recruit iron workers. While the political situation in England was unsettled, he visited alchemical friends in Europe and honed his iatrochemical skills. There was a vibrant intellectual world in England and the Colonies inspired by Francis Bacon and instantiated in the 1640s by Samuel Hartlib. Winthrop was a full member of this world. Their goal was a Protestant utopia enlightened by a direct interaction with the real world. The story of the production of iron in New England is presented in some detail, including the chemistry.

While small scale activities can be carried out with modest means, genuinely industrial production requires substantial capital. In 1641 John Winthrop, Jr., returned to England with a goal of capitalizing an iron works. While initial enthusiasm predicted a short visit, Winthrop did not return to New England until 1643. The political side of his visit was carried out in collaboration with Hugh Peter and Thomas Weld (1595–1661). Peter and Weld both remained in England and played large roles in the Parliamentarian government. They never achieved their goal of receiving an extension of the charter for Massachusetts. Another scheme of Peter was to subvert the Rhode Island Colony founded by Roger Williams, but this also failed. When they arrived in England, the Civil War made it difficult to carry on any kind of advocacy for a far-off place like New England. Rather than merely wasting time, John Winthrop, Jr., visited Europe and experienced a major change in his life.

There were two contacts for John Winthrop, Jr., that opened up the European world of alchemy: (1) Samuel Hartlib in London had the most extensive circle of correspondents, and (2) the brothers Abraham and Johann Kuffler from Amsterdam knew most of the alchemists and alchemical physicians in Europe. John Winthrop, Jr., met the German physician, Johannes Tanckmarus in Hamburg. Germany had been home to the famous iatrochemists, Oswald Croll (1560–1608), Johannes Hartmann (1568–1631) and Andreas Libavius (1540–1616). Winthrop learned a great deal about Paracelsian medicine and determined to bring this benefit to New England. Later he met Rev. Johann Moraien (1591–1668), another alchemical physician in Amsterdam who held public clinics and prescribed alchemical remedies either for

© The Author(s), under exclusive license to Springer Nature Switzerland AG 2020   43
G. Patterson, *Chemistry in 17th-Century New England*,
SpringerBriefs in History of Chemistry,
https://doi.org/10.1007/978-3-030-43261-4_6

free or a small donation [1]. While in Amsterdam, Winthrop also met the most famous chemist of the time, Johann Rudolph Glauber (1604–1670) [2]. John Winthrop, Jr., created the finest library in America in the seventeenth century and it included many of the more than 40 books written by Glauber. Another important contact while in Holland was the Walloon ironmakers. Their "indirect" process was much better than a raw "bloomer" and was used at the ironworks founded by Winthrop at Saugus, Massachusetts [3]. Moraien also introduced him to a Dutch investor, John Becx, who joined the Company of Undertakers of the Ironworks in New England.

In antiquity, the production of iron from the ore was carried out at a temperature below the melting point of pure iron and the result was a mixed mass called a "bloom." The furnace was primitive and consisted of a truncated conical structure made of bricks with a hole in the top and at the bottom. There were also air holes that could be regulated. The top hole received the initial charge of charcoal and the furnace was fired until the proper temperature was achieved. The bog iron ore was separately preheated to drive off any excess water included in the ore. The next charge included both the bog iron ore and enough charcoal to fully reduce the iron ore. The two components were intimately mixed. As the charcoal was oxidized to carbon monoxide it reacted with the iron oxide and produced small pieces of iron. This released more heat, but not enough to melt the iron. The other components of the raw ore were melted and the final product of the furnace was a bloom containing iron, charcoal and "slag." This highly heterogeneous mass was allowed to cool and was removed by opening the hole in the bottom part of the cone. It was called "sponge iron." The hunk was heated and hammered (wrought) until the slag pieces were removed and the excess carbon was oxidized. Bloomery iron was produced in Colonial America later than the Winthrop effort, but he had higher ambitions.

Upon his return to England, Winthrop settled in London and began his solicitation for investment in an ironmaking company: The Company of Undertakers of the Ironworks of New England. This effort was highly successful, and John Winthrop, Jr., put in motion a plan to recruit workers and buy capital equipment. He also secured shipping space for the journey to New England. His primary purpose in coming to England was a success.

But, since he was in London, he sought out kindred spirits. The most important person that he met and developed a long-term friendship with was Samuel Hartlib (1699–1763) [4]. Hartlib was the center of a "circle" of scientists and religionists who sought universal Protestant religion and the "Advancement of Knowledge." Hartlib was inspired by the Moravian bishop, Jan Amos Comenius, and Winthrop met and became a disciple of this charismatic figure. While John Winthrop, Jr., remained an official Puritan, his private convictions were influenced by the desire to achieve the Protestant millennial utopia in New England. The Hartlib circle also included people like Sir Christopher Wren, John Milton, Sir Kenelm Digby, Benjamin Worsley [5] and Robert Boyle [6]. The friends that Winthrop made on this trip to London were a great help to him for the rest of his career.

6.1 Francis Bacon and the Great Instauration

While Francis Bacon (1561–1626) was hardly a Puritan, he promoted ideas that were at the heart of the Puritan project. He believed that God was about to help Protestant mankind to achieve a great increase in learning and become the masters of the physical world [7]. While "modern" science often proceeds entirely without reference to the Providence of God, many European alchemists of the seventeenth century prayed daily for illumination directly from God. John Winthrop, Jr., was wholly committed to this stance.

Another key element in the Baconian programme was the close coupling between empirically derived knowledge and its application to the needs of humanity. The notion of "knowledge for knowledge sake" was quite foreign to the Puritan worthies of the seventeenth century. However, fundamental investigations of the properties of matter, both then and now, needed to be carried out with surplus money. Governments insisted on financial rewards for money invested in "applied research." John Winthrop, Jr., engaged in both kinds of research. Since the arcane knowledge he received was viewed as from God himself, he was careful to protect it from the general public. Rather than being merely a "trade secret," he viewed it as a sacred trust.

Another aspect of the Baconian paradigm was the primacy of actual observation over mere speculation. The philosophical battle between a priori speculation and empirically grounded inductions is still raging, but in the seventeenth century the preference was for theoretical explanations grounded in good data. John Winthrop, Jr., "worshipped" at the "Altar of Vulcan." He carried out many alchemical experiments at his furnace and regularly synthesized the medicines he used in his practice. He chose not to create elaborate literary fictions like his contemporary George Starkey [8].

6.2 Ironworks at Braintree

An outstanding discussion of iron-making in New England is contained in the monograph, *Ironworks on the Saugus* (1957), by Hartley [3]. John Smith had evaluated this area on his voyage of 1614 and concluded "that anyone who would set up a forge where food and ore and charcoal were free could not lose [9]."

John Winthrop, Jr., had carried out a more thorough survey of greater Boston and also concluded that there was enough bog iron (Fig. 6.1), trees and water power to make an ironworks a profitable venture.

Iron is one of the most common minerals on the surface of the Earth and constitutes the dominant constituent of the subterranean core. Streams and springs harvest iron from the ground and often deposit it in bogs. That might be the end of the story, except that *bacteria* in the bogs ingest the iron and eventually create bog iron nodules. The

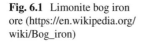

Fig. 6.1 Limonite bog iron
ore (https://en.wikipedia.org/
wiki/Bog_iron)

oxidized form of iron is typically $FeO(OH) \cdot n(H_2O)$. There are indeed many bogs in Massachusetts. They are now used primarily for the cultivation of cranberries!

When John Winthrop, Jr., had returned to Massachusetts, he needed to make the local arrangements for his ironworks. First, he needed to choose a suitable site with good water power, plenty of trees and access to both housing and transportation. He chose Braintree, Massachusetts, a town on the Fore River, ten miles due south of Boston. While the site appeared to be ideal, the land was already under private ownership. Winthrop chose to purchase the land in order to secure full rights to build and operate the works. Then he needed to secure monopoly rights from the Massachusetts Bay Company, which he easily obtained as an Assistant. In addition, he secured freedom from taxation and freedom from militia duty for the workers. The one stipulation that the General Court made was that no iron could be shipped outside New England until all the local needs were met. Actual industrial chemistry depends as much on external factors as on the details of the chemical process. While his workers knew something about the production of iron, they were neither Puritans nor particularly "godly."

With his English capital in hand and an English workforce in place, he constructed the plant and fired up the furnace. There were plenty of trees and the production of charcoal was an established art. The reduction of actual trees to usable charcoal, while known for millennia, is largely unknown among the general populace today. The final product is essentially carbon in a highly porous, yet mechanically robust, form. Trees are harvested in the winter when the sap is in the roots. The bark is stripped off and used for mulch. The logs are cut into four foot lengths for easy stacking. The creation of a good charcoal stack is an art passed from one generation to another of "coalers" (Fig. 6.2).

The two key concepts in the production of charcoal from wood are: (1) Dehydration of the cellulose, and (2) Limited combustion to provide enough heat without destroying the carbon. To accomplish this, the pile is conical and a hole in the peak is created at the start. There are also air holes at the bottom around the sides that can be closed when needed. The wood pile is covered with an insulating layer that excludes air. In the seventeenth century, this was usually sod. Later practice was to

Fig. 6.2 Charcoal stacks in various stages of construction [public domain image attributed to Diderot (1763)]

use bricks or stones to make permanent, reusable "kilns." Once the fire was started, the smoke was examined to determine the extent of the process. The fire smoldered in the initial stages and the smoke was white. When dehydration was more complete, the smoke changed to "blue" as the temperature rose. Once the process was finished, the holes were plugged to stop combustion and the pile was allowed to cool. The charcoal cylinders were solid, but fragile, and great care was taken to maintain the physical shape of the carbon rods. Good "charcoal rakers" were highly valued.

The pure chemistry is very simple:

$$CH_2O(\text{cellulose}) \rightarrow C(c, gr) + H_2O(g)$$

but the actual technology is tricky and failures were common. If the pile got too hot, it might consume all the wood. If the final logs were broken or, worse, pulverized, the final charcoal would be unsuitable for the ironmaking process then in use. Fortunately, Winthrop hired good Scottish "slave labor" to make the charcoal, under the direction of expert coalers.

Reduction of the bog iron to pig iron involved mixing the charcoal with the rocks. Just heating this in a vacuum would have produced nothing. Again, a fire was started in the charcoal, and the burning conditions were very spare on air. When the proper conditions were obtained, the charcoal was oxidized to carbon monoxide.

$$C(s, gr) + (1/2)O_2(g) \rightarrow CO(g)$$

The gaseous CO could interact with the hot, porous bog iron to reduce the iron and produce carbon dioxide and water.

$$2FeO(OH)(s) + 3CO(g) \rightarrow Fe(l) + 3CO_2(g) + H_2O(g)$$

The molten iron was collected from the bottom of the furnace. The oxygen concentration was carefully controlled to keep a reducing atmosphere. The reduction itself is exothermic so that once ignition is obtained, only enough oxygen is needed to produce the carbon monoxide.

While the "pure chemistry" is very simple, industrial chemistry must include all the issues needed to produce the desired product. Real iron ores contain many impurities, and they must be separated from the pure liquid iron. The secret is to add a "flux" that will also be liquid, not mix with the molten iron, and either dissolve or react with the impurities. Limestone was the most common flux after it had been "calcined" to CaO(s). A drawing of a "blast furnace" is shown in Fig. 6.3.

The "blast" of air delivered to the molten iron oxidized the remaining carbon fragments to produce a purer liquid iron. It was also more dense than the flux and could be drained out from the bottom of the furnace. Industrial chemistry accounts for all the actual processes involved. The furnace is highly insulated with both inner and outer walls. Since the temperatures were very high, the inner walls needed to be lined with "fire brick." The heat needed to start the process was obtained from partial burning of the charcoal. This is an explicitly heterogeneous process, with gases, liquids and solids. Transport of both energy and materials must be facilitated. The "pure" iron must be separated from the "slag" at the end of the process. Separation technology is one of the characteristic "unit operations" of industrial chemistry. None of these things were being taught in universities; they were all "craft skills" that were taught within the guild community. John Winthrop, Jr., brought them to New England.

After everything was ready to go, he had exhausted his capital and appealed to England for more. They promptly sacked him and replaced him with Richard Leader (1609–1661), an experienced iron-maker. It was one thing to envision an ironworks, and another to actually construct and run one at a profit. Nevertheless, iron was actually produced at Braintree for over a century. Leader surveyed the New England landscape for a better site and selected Saugus, Massachusetts. The complete ironworks, including charcoal, furnace, and wrought iron and slitting mills, was called Hammersmith. A professional reconstruction is illustrated in Fig. 6.4.

John Winthrop, Jr., remained interested in the ironworks and helped to obtain the needed grants of land, water rights, and monopoly protection. The wrought iron

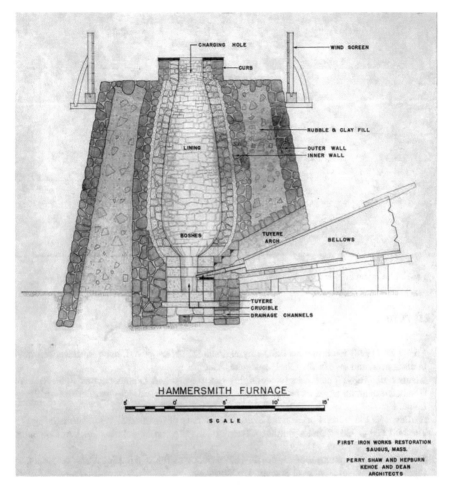

Fig. 6.3 A drawing of the reconstructed blast furnace at Saugus, MA (National Park Service, by permission)

forge, the slitting mill and other machine shops required power, and water wheels were the method of choice in New England. Leader was a master at this technology and went on to found saw mills as well.

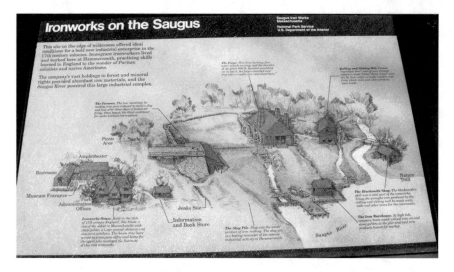

Fig. 6.4 The ironworks at Saugus Massachusetts (National Park Service, by permission)

References

1. Young JT (1998) Faith, medical alchemy and natural philosophy: Johann Moriaen, reformed intelligencer, and the Hartlib Circle. Ashgate, Aldershot
2. Glauber JR (1689) The works of the highly experienced and famous chymist John Rudolph Glauber: containing great variety of choice secrets in medicine and alchymy. Christopher Packe, London
3. Hartley EN (1957) Ironworks on the Saugus. University of Oklahoma Press, Norman
4. Turnbull GH (1920) Samuel Hartlib: a sketch of his life and his relations to J.A. Comenius. Oxford University Press, Oxford
5. Leng T (2008) Benjamin Worsley (1618–1677): trade, interest and the spirit. In: Revolutionary England. The Royal Historical Society, Woodbridge
6. Hunter M (2009) Boyle: between god and science. Yale University Press, London
7. Gaukroger S (2001) Francis Bacon and the transformation of early-modern philosophy. Cambridge University Press, Cambridge
8. Newman WR (1994) Gehennical fire: the lives of George Starkey, an American alchemist in the scientific revolution. The University of Chicago Press, Chicago
9. Smith J (1907) The Generall Historie of Virginia, New England and the Summer Isles. Macmillan, New York

Chapter 7
New London and the Alchemical Plantation

Abstract After John Winthrop, Jr., had been relieved of his responsibilities for the ironworks, he was free to pursue his dream of creating a utopian plantation in Connecticut. This is a complicated story, but it illustrates all the excellencies of Winthrop. He founded a flourishing town of New London. He established an iatrochemical practice that all the leading citizens of Connecticut used. He negotiated the difficult political situation with both the Colonial governments and the Native American tribes. And he practiced alchemy and industrial chemistry. It is no real surprise that he was elected Governor of Connecticut without running for office. One of his favorite medicines was based on antimony. The chemistry of this element is described. Another important chemical was saltpeter, and Winthrop manufactured it as well.

John Winthrop, Jr., had not forgotten about his plans for a plantation at Saybrook on the Connecticut River, and while in the area on other business, he visited sites in the surrounding countryside. He also explored sites near what became New London and decided to found a plantation there in the Pequot (now Thames) River Valley [1]. While in England he had also obtained title to Fischer's Island, off the coast of Connecticut near the mouth of the Pequot. During 1645–1646 he had a house built on this island and occupied it during the winter of 1646.

Since Winthrop already had multiple residential and business interests, what could have motivated him to launch an even more adventurous project? The story starts in the early seventeenth century with many people promoting the idea of an alchemical enterprise composed of the best natural philosophers in the world. Francis Bacon's utopian novel *The New Atlantis* was published posthumously in 1627 by his literary executor, Dr. Rawley. It presents an institution, *Solomon's House,* comprised of the best natural philosophers dedicated to discovering the ways of nature for the benefit of mankind. It is highly likely that Winthrop read this work, perhaps even in manuscript. An even earlier work, the *Republicae Christianopolitanae Descriptio* (1619) by Johann Valentin Andrae, described the utopian city of Christianopolis. It is inhabited by citizens committed to the joint pursuit of truth and practical righteousness. The workers are both skilled artisans and natural philosophers. Andrae was the author of the celebrated Rosicrucian Manifesto that so excited Edward Howe

G. Patterson, *Chemistry in 17th-Century New England*,
SpringerBriefs in History of Chemistry,
https://doi.org/10.1007/978-3-030-43261-4_7

and John Winthrop, Jr. [2]. The contemporary motivation for an alchemical plantation was promoted by a collaborator of Hartlib, Gabriel Plattes (1600–1644), in a book *A Description of the Famous Kingdome of Macaria; shewing its excellent Government: wherein the Inhabitants live in great Prosperity, Health, and Happiness* (1641) [3]. Plattes was especially interested in "improved husbandry" and Winthrop followed many of his ideas on Fishers Island. He was also a dedicated chemist and friend of Robert Child. If Plattes had lived, perhaps he would have joined Winthrop in Connecticut. They met during Winthrop's time in London in 1642–1643. One of the central concepts in *Macaria* was governmental support for research in practical applications of the universal knowledge obtained by the natural philosophers. The high hopes in the Protestant Parliamentarians to instantiate this "True Reformation" of society was dashed by the realities of the English Revolution.

In order to support the "research" at the plantation, it was necessary to make money by other means. Husbandry and agriculture were good candidates to provide the cash flow. Another possibility was mining. Winthrop had been given samples of "black lead" gathered from a "mine" at Tantiusque, a site at the headwaters of the Pequot River in Massachusetts. Assays of the ore differed while he was in England, but the final conclusion was that it was graphite (called plumbago in the seventeenth century); not exactly a "precious metal" but still valuable nonetheless. John Winthrop, Jr. purchased the mine site from two different groups of Native Americans in order to forestall any competition once he started to harvest the ore. Nevertheless, the realities of such an industrial venture in seventeenth century New England were unfavorable. Even investment by Robert Child and actual mining by Thomas King did not bring the project to success.

In the seventeenth century, what we now know as graphite was called either "black lead" or *plumbago*. It was used as the writing element in "pencils" and in pure rods by artists and shepherds (to mark the sheep). An example of "black lead" ore is shown in Fig. 7.1.

The chosen site on the Pequot River was perfect from a geographical perspective, but highly fraught from a political point of view. The Native Americans in this part of New England were all from the "Algonquian" group, but every tribe followed its own interests. The leader of the tribe was called a Sagamore or a Sachem, and a warlike chief could either ruin the tribe or obtain great glory. There was an Indian settlement called Nameaug at the site of the proposed plantation that was associated with the Pequot tribe. A serious war between the Pequots, Narragansetts and Mohegans concluded in 1638 with the Treaty of Hartford. This document called for the obliteration of the Pequot tribe, with the former members being distributed as slaves among the other two tribes, and the Commonwealths of Massachusetts and Connecticut. Active Pequot warriors were executed. Noncombatant males were enslaved or subjected to punitive tribute. Once the graves had been dug, the fight for the former Pequot land started. John Winthrop, Jr., obtained a grant from Massachusetts to found a plantation at Nameaug in 1644, and started laying out the plantation in 1645, along with the former minister from Saybrook, Thomas Peter (the brother of his father-in-law Hugh Peter) [4]. Winthrop wintered with his family on Fishers Island, where he had a house.

Fig. 7.1 Mineral graphite
("black lead")

The local Sachem was a Mohegan chief named Uncas. He made normal life precarious for Pequots and Englishmen in his territory. He also eyed the Narragansett lands and people. The Narrangansett sachem, Miantonomo, launched a pre-emptive attack, but was captured by Uncas. Rather than execute him on the spot, Uncas turned him over to the Connecticut authorities for justice. Under the cover of English justice, Miantonomo was "released" so that he could be assassinated by Mohegan warriors. The Narragansetts and the Mohegans continued their agitations as long as Winthrop was alive. Soon after John Winthrop, Jr., and Thomas Peter arrived to survey the site for New London, the Narragansetts tried to exact revenge on the Mohegans. Many were wounded, including Uncas. John Winthrop, Jr., and Thomas Peter went to the Mohegan fort of Shantok and rendered medical treatment, saving many lives. One might imagine that this would produce a sense of gratitude on the part of Uncas, and his initial response was positive. But, a year later Uncas and 300 Mohegan warriors conducted a raid on Nameaug and killed some, robbed all and burned the wigwams while John Winthrop, Jr., was away in Boston. When he returned, the English settlers appealed to the government of Connecticut, but were disappointed when the magistrates concluded Uncas was merely exerting his rightful rule over

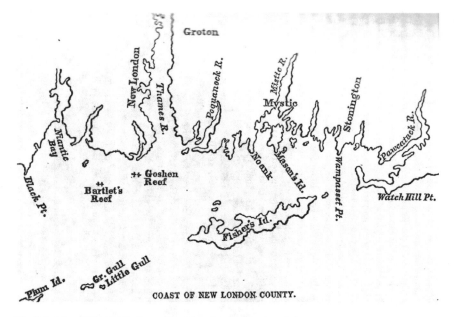

Fig. 7.2 Map of Connecticut near New London (public domain)

the Pequots at Nameaug. And if the English had the poor judgement to live there, so much the worse for them! But, in the event, the classic *History of New London* (1895) by Frances Manwaring Caulkins bears witness to the perseverance of John Winthrop, Jr., and the brave souls that founded the town [4].

John Winthrop, the Younger, had experience in founding a town, and the initial stages of settlement of New London followed sound principles. The town was well surveyed and each early settlor received a town lot, a woodlot, and a meadow. In addition, land was reserved for common use, church use and municipal use. In addition, inducements to enterprise were given to craftsmen like millers, blacksmiths and clergymen. What New London had to offer was land, and John Winthrop, Jr. accumulated a squirearchical amount. He also received civic help in creating a mill pond and water race to drive a saw mill and a gristmill. He constructed a fine stone house on a promontory overlooking the West bank of the river. The town founded on the East bank of the Pequot was named Groton, in honor of his English estate. In many respects it could have been Brigadoon, but no town can actually remain an island unto itself (Fig. 7.2).

Once the town was well-founded and both food and shelter were assured, John Winthrop, Jr., set about the business of alchemical medicine. He constructed his furnaces and started to produce the antimony medicines so beloved of Johann Moriaen [5]. All the leading families of New England resorted to New London for serious illnesses and Winthrop provided both medicines and hospice until all was well. In the finest alchemical tradition, he did this as a service to God. The patients were

Fig. 7.3 Sample of mineral
stibnite showing the
columnar nature of the
crystals

treated individually and conservatively. Isolation, careful hygiene, good nutrition, and personal observation did wonders. But, he could not cure smallpox!

7.1 Antimony

Stibnite (Sb_2S_3) has been known since antiquity as a finely powdered substance called kohl, and was widely used in Egypt as a cosmetic (Fig. 7.3).

Reduction of stibnite to metallic antimony was achieved by heating the ore with metallic iron. The separated metallic antimony is not quite pure but produces a beautiful form called the star regulus of antimony (Fig. 7.4).

The form used by John Winthrop, Jr. was the trioxide (Sb_2O_3). It was often called diaphoretic antimony, since it produced perspiration. It could be produced from either stibnite or the regulus by roasting in air and purifying either by sublimation or by deflagration with saltpeter (KNO_3).

7.2 Saltpeter

One of the most valuable chemical substances needed for success in a Colonial setting was saltpeter. It was one of the essential ingredients in gunpowder, a common ingredient in medicine and a fertilizer. Purchasing it from Europe was expensive. John Winthrop, Jr., was very interested in producing saltpeter on Fisher's Island. There was an ample supply of hardwood branches and leaves needed to make "potash" (K_2CO_3).

Fig. 7.4 Star regulus of
antimony, a polycrystalline
material (produced by the
Chemistry of Isaac Newton
Project at Indiana University,
by permission)

He raised goats and sheep on the island; this produced lots of good excrement and urine. There is plenty of nitrogen in the atmosphere, but in the seventeenth century, this was hardly known. The key step was oxidizing the nitrogen to nitrate. The animal manure, which contained both feces and urine, was a good source of nitrogen as ammonia (NH_3). How could this be transformed into nitrate? As with many actual chemical processes in antiquity, biological organisms were the key producers. Aged human "excrement" was harvested by force in England (and many other countries) and "bat guano" was harvested on isolated islands. But, the colonists needed to produce saltpeter much more quickly and cheaply than that.

The basic recipe used by the English for making saltpeter was purchased from Gerard Honricke in 1561 [6, 7]. Artisanal manufacture of saltpeter was common during the Middle Ages, and Natural Philosophers thought that they could easily improve on the process and make it better and more cheaply. Why did they, and John Winthrop, Jr., fail? The root of the problem was the difference between artisanal trade secrets and faulty academic notions of the nature of matter. In the seventeenth century, there was no real understanding of exactly what saltpeter was. It was obtained from animal and vegetable sources, but it behaved like a typical mineral salt. Johann Glauber published both recipes for its production and speculations about its properties and nature [8]. Glauber focused on the virtues of salts, and attributed almost miraculous powers to "aerial nitre." With this level of confusion, even among the best alchemists of the century, it is little wonder that the observed results for the production of saltpeter were inconsistent.

Because of the use of animal excrement in the process, it was assumed that this was the "starting material" in the process. But, careful modern experiments have demonstrated that pure manure yields no saltpeter by itself [7]. The actual reactant at the start of the process is urea, obtained from urine (Fig. 7.5).

Fig. 7.5 Structural and space-filling diagrams for urea (CON_2H_4)

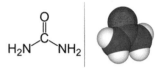

This readily available chemical is far from the nitrate (NO_3^-) needed for saltpeter. What does the manure contribute to the process? Bacteria! It is now known that *Nitrosomonas* and *Nitrobacter* are required to oxidize the ammonia (NH_3) produced by hydrolysis of urea to nitrate. Oxidation requires oxygen, so that the "nitre bed" needed to facilitate the transport of atmospheric gases into the bed. This was achieved by adding good humus and frequent "turning" of the beds. The soil also was a source of bacteria. Saltpeter manufacture was a form of biotechnology, but since none of the concepts necessary to understand this existed in the seventeenth century, many attempts to produce "fertile" nitre beds failed.

Another reason for the inconsistent results was a lack of patience. Traditional "saltpetremen" only harvested truly aged "nitrous earth" which "tasted right." Projectors like Benjamin Worsley were "in a hurry" to produce the saltpeter and did not wait long enough for the process to go to completion. Alchemical speculations suggested that the Earth would "produce" all the materials needed by God's creatures, if either the materials were readily apparent or *adepts* knew how to assist Nature in her work. Artisanal producers did know that good "nitre beds" should be re-used, just like "starters" for bread or beer. No bacteria, no saltpeter! Naturally occurring saltpeter "efflorescence" was observed on walls that had been "blessed" with both dung and urine over extended periods of time. Standard "mortars" in the seventeenth century included dried manure in the mix, and the addition of urine provided both water and urea.

The metallic constituent of saltpeter was potassium, but this element had not yet been "discovered" in the seventeenth century. Nevertheless, artisanal producers used "potash" (K_2CO_3) in the process. In spite of the lack of detailed knowledge about the chemical nature of potash, it had been used in artisanal chemical industry for millennia.

Industrial quantities of saltpeter could not be obtained from "wall scrapings." The processing of nitre beds was tedious and uncertain. The basic idea was to extract the "active salt" from the solid mixture by subjecting it to water (*lixiviation*). The mixing of the water and the dirt was tricky, and many mud pies yielded very little active ingredient. Draining of the solution from the slurry often led to plugged hoses. But, artisanal experts knew how to deal with the ever-changing outcomes. Patience was rewarded, while haste was punished. Once the solution had been harvested, concentration and purification were needed. This was typically done by boiling the solution and skimming the impurities from the surface. Also, precipitated impurities were separated by decanting the supernatant liquid. Many vessels were required,

Fig. 7.6 Saltpeter crystals

and since heating was employed, copper kettles were the best choice. Wooden tubs tended to burn or leak. This level of industrial expertise was often not revealed in the "recipes."

If things went well, a solution of calcium nitrate would be obtained. The calcium was obtained from the manure, the soil and the local water. The solution was now filtered through beds of potash to produce a double exchange of constituents. Calcium carbonate precipitated from solution and the final solution contained potassium and nitrate, in addition to other salt constituents such as sodium and chloride. The final filtered solution was concentrated until it was saturated with saltpeter and then cooled to produce the crystals (Fig. 7.6).

Common salt remained in solution. During the seventeenth century, saltpeter produced according to the artisanal practice was rarely pure and sparkling white. But, it was good enough for gunpowder. John Winthrop, Jr., needed purer materials to use as a medicine, and further steps were employed to produce medicinal grade saltpeter. Reprecipitation and filtration from solution was used.

7.3 The Leading Citizen of Connecticut

Colonial governance was a very fluid thing. The "Puritan" plantations, Massachusetts, Plymouth, Connecticut and New Haven, ignored Providence Plantation (Rhode Island). However, John Winthrop, Jr., maintained good relationships with Roger Williams, the Governor of Rhode Island, and this allowed him to navigate the treacherous waters of Native American culture. When Winthrop established his New London Plantation, it was not clear which Colonial entity actually governed it. In the end, it was attached to Connecticut, but John Winthrop, Jr., made sure to cooperate with the full range of Puritan governments. As late as 1648 Winthrop received a grant of 3000 acres near the mouth of the Pawcatuck River (see Fig. 7.2) from the Massachusetts Bay Company, of which he was still an Assistant. This site was intended as another saltworks. His full detachment from Massachusetts occurred after the death of his father in 1649.

John Winthrop, Jr. was admitted as a "freeman" of Connecticut in 1650. He also officially changed his church membership from Boston to Saybrook. He consolidated his activities at New London and established enterprises such as a mill. His farm on Fisher's Island continued to produce sheep and cattle. Not even the Native American policy of stealing his goats hindered his efforts. He created a stock company for a saltpeter works at New London. From his excellent location on the Connecticut coast, he was in contact with American settlers from Maine to Delaware, and established a close relationship with Peter Stuyvesant, Governor of New Amsterdam.

While New Haven was a separate Colony, Puritan connections were strong and regular communication occurred. John Winthrop, Jr., offered more than any other person to a struggling town in Colonial New England. Reverend John Davenport, the leading citizen of New Haven, had consulted him for his medical expertise in 1653. With the success of the Roundhead Revolution in England, many Puritans in New Haven considered returning to the motherland. New Haven had developed very little industry and was eager to establish an economically sustainable community. The town fathers of New Haven actively pursued John Winthrop, Jr., and offered him a fine house and servants. They hoped he would found an ironworks and other enterprises. There were many bogs in New Haven Colony. A company was formed with John Davenport, Theophilus Eaton, Stephen Goodyear, Jasper Crane and John Winthrop, Jr., to build an ironworks and the furnace was fired in 1657. However, actual production at New Haven was not successful until 1663. In reality, it would have been difficult for John Winthrop, Jr., to thrive in the ultra-Puritan culture of New Haven.

The efforts of Connecticut to retain the services of John Winthrop, Jr., led to his election as Governor in 1657. He moved to Hartford, and served the colony until his death. Nevertheless, he retained his property at New London and its many enterprises. He was an astute politician who tried to reach amicable solutions to problems. He even managed to help spare the life of an accused witch. It was his expertise as a chemical adept that gave him considerable authority in "occult matters." In the seventeenth century, alchemy was as much a "spiritual discipline" as it was an artisanal practice, and John Winthrop, Jr., used his status as an adept to do good [9]. Although the Connecticut charter required that the Governor not succeed himself in office, John Winthrop, Jr., was re-elected every year until his death!

References

1. Black RC (1966) The Younger John Winthrop. Columbia University Press, New York
2. Yates FA (1972) The rosicrucian enlightenment. Routledge and Kegan Paul, New York
3. Plattes G (1641) A description of the famous kingdom of Macaria. London
4. Caulkins FM (1895) History of New London. Connecticut H.D. Utley, New London
5. Young JT (1998) Faith, medical alchemy and natural philosophy: Johann Moriaen, reformed intelligencer, and the Hartlib Circle. Ashgate, Aldershot
6. Williams AR (1975) The production of saltpetre in the middle ages. Ambix 22:125–133
7. Robertson H (2016) Reworking seventeenth-century saltpetre. Ambix 63:145–161
8. Glauber JR (1689) The works of the highly experienced and famous chymist John Rudolph Glauber: containing great variety of choice secrets in medicine and alchymy. Packe C, London
9. Woodward WW (2010) Prospero's America: John Winthrop. Jr., alchemy, and the creation of New England culture, 1606–1676. University of North Carolina Press, Chapel Hill

Chapter 8
Obtaining the Charter for Connecticut and Election to the Royal Society of London

Abstract After the restoration of Charles II, John Winthrop, the Younger, traveled to London to obtain a Royal Charter for Connecticut. While it took a long time and required a large sum of money, Winthrop showed both political skill and patience in obtaining his desired end. He also used his copious free time in London to pursue his scientific interests among the Hartib Circle. He even became a founding member of the Royal Society. A brief discussion of the contributions to chemistry in 17th century New England by Cotton Mather is also presented.

Once the English Revolution ran its course and Charles II was restored to the throne, wise Americans needed to make peace with the Crown. John Winthrop, the Younger, was ideally suited for this task. He had powerful friends in England and had kept a low profile during the Interregnum. It was now time to obtain a Royal Charter for Connecticut. In order to do this, John Winthrop, Jr., needed to navigate treacherous political waters and also make a trip to England. After careful preparations in New England, he chose to sail from New York, with the knowledge and assistance of Peter Stuyvesant, the Governor of New Amsterdam. He arrived in Continental Amsterdam on September 6, 1661. After pleasantries with his Dutch friends there, he proceeded to The Hague, where his cousin, George Downing, was the English Ambassador. He finally reached London on September 18, 1661. He set up shop in the house of William Whiting, a son of a former treasurer of Connecticut. This was a perfect place to forge the Puritan campaign for a charter [1].

He soon called on his old friend Samuel Hartlib. While he did not have any influence in the court of James II, he did know a great deal about the current situation. Hartlib also recommended John Winthrop, Jr., to Benjamin Worsley. Worsley was on good terms with the Lord Chancellor, Clarendon, and was connected with the Privy Council on Trade and Plantations. He was a Natural Philosopher and Natural Historian. He was also a friend of Robert Boyle, who was also on the Council of Plantations. Worsley was able to assist Winthrop with many issues. A common interest in saltpeter fostered other interests!

The days waiting for the opportunity to present his petition before the King were not wasted [2]. He established scientific and personal relationships with many of the founders of the Royal Society: Baron William Brereton (1631–1680), Sir Robert

G. Patterson, *Chemistry in 17th-Century New England*,
SpringerBriefs in History of Chemistry,
https://doi.org/10.1007/978-3-030-43261-4_8

Moray (1609–1673), Henry Oldenburg (1619–1677), Hon. Robert Boyle (1627–1691), Elias Ashmole (1617–1692), Rev. Dr. John Wilkins (1614–1672), Sir Isaac Newton (1643–1727), John Milton (1608–1674) and Christopher Wren (1632–1723). It was Brereton who first brought John Winthrop, Jr., to an early meeting at Gresham College. He was proposed as a candidate on December 18, 1661 and admitted to membership on January 1, 1662. His name appears in the list of Original Fellows in 1663.

John Winthrop, Jr., was able to present his petition for a charter for Connecticut to the King on February 12, 1662. The advice and personal intercession of Sir Robert Moray was very helpful. It was accepted and passed on to the Attorney General, Sir Geoffrey Palmer, to advise on the details of the charter. With remarkable speed, the initial draft of the charter was approved on February 28, and the final form passed the Privy Seal on May 10. Nevertheless, it would be many months before all the machinations associated with the process were settled and John Winthrop, Jr., could return to Connecticut. Fortunately, Robert Boyle interceded and introduced him to Henry Ashurst (1614–1680). Ashurst was both wealthy and interested in the Corporation for Propagating the Gospel in New England. The money needed to protect the charter from attack and to provide living expenses was provided.

During his extended stay in London, John Winthrop, Jr., became not only a regular at meetings of the Royal Society, but participated enthusiastically in its activities. He was appointed to committees for the history of trade and for mechanical inventions. He exhibited many "rarities" from New England, a favorite topic in the early Royal Society. Winthrop discoursed on the growing and eating of New England maize, showed a rattlesnake tail and shared mineral specimens.

One of his most notable contributions was both a presentation and a finished article on the process for making "tar" from pine trees in New England. It appears also as a section of the volume *Sylva: Or a Discourse of Forest Trees* by John Evelyn (1620–1706) FRS, the famous arborist [3]. (Evelyn is remembered today for his *Diary*.) Parts of New England are littered with "the knots of pitch pines." These are filled with *Terebinthin* which can be extracted by heating in a sealed oven and drawn off after the process is finished. Pine tar was one of the most important substances in the ship building industry, since it served as a preservative for wood. A diagram of a pine tar oven is shown in Fig. 8.1. A natural slope provided one wall of the oven. The liquid oleoresin was collected using a stone base and a barrel. The heat was provided from above by using "brushwood."

The starting material, pine knots, was an adventitious product of both natural forces and Native American agricultural practices: burning the forests. They were collected from the ground and did not rot, since they were saturated with pine tar. At the end of the extraction process, the knots were reduced to highly valuable charcoal. Even in Colonial times, wise chemists thought like chemical engineers: (1) They obtained inexpensive and readily available starting material. (2) They used scrap materials as the source of energy. (3) They maximized the production of useful "product." (4) They recovered valuable "by-products." John Winthrop, Jr., understood the need to make the most of his environment, both chemically and politically.

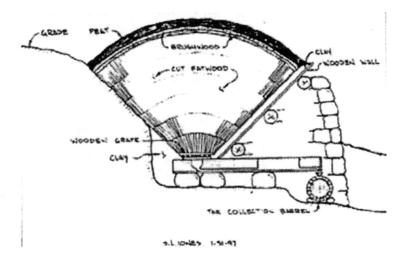

Fig. 8.1   A pine tar extraction oven

Another one of his notable discourses was on the nature and uses of Indian corn (maize). While the science of agriculture was in its infancy, Native Americans had developed both early and late corn varieties. Winthrop considered the multitude of substances that could be obtained from the raw plant. He anticipated the importance of "corn syrup!" While ignorant English connoisseurs denigrated the nutritional value of corn meal, Winthrop treated them to the pudding, called "Sampe," that was actually consumed in America. Baked cornbread could be eaten directly, but it had another important use: it was the basis of a type of "middle beer." Winthrop brewed some and served it to the Royal Society. Only a lack of time prevented him from serving the first Colonial corn whiskey!

John Winthrop, Jr., was a good example of a member of the Royal Society. Rather than becoming an expert in a very narrow specialty, he stayed alert to the bigger issues of life and economy. Rather than sticking to one area of Natural History, he viewed animal, vegetable and mineral kingdoms as connected in essential ways. The connection between the vegetable world and the world of terrestrial charcoal was understood. While it was still too soon to understand the role of true microorganisms in the production of both oxidized and reduced forms of matter, he trusted the effective recipes that mixed known ingredients to produce valuable products.

After his return to Connecticut, he set about collecting new rarities to be sent to London. But, just as now, safe delivery was not guaranteed, or even likely. A large shipment of novelties sent in 1665 was "lost at sea [2]." Henry Oldenburg wrote to Winthrop encouraging him to remember his Fellows in London [4] (Fig. 8.2).

Finally, in 1669, he sent a very large collection of Americana to England in the company of Adam Winthrop (1645–1700). "The shipment consisted of four boxes, including upwards of 50 specimens, groups of specimens, and two copies of Eliot's Indian Bible, an Indian grammar, works of piety translated into the Indian language,

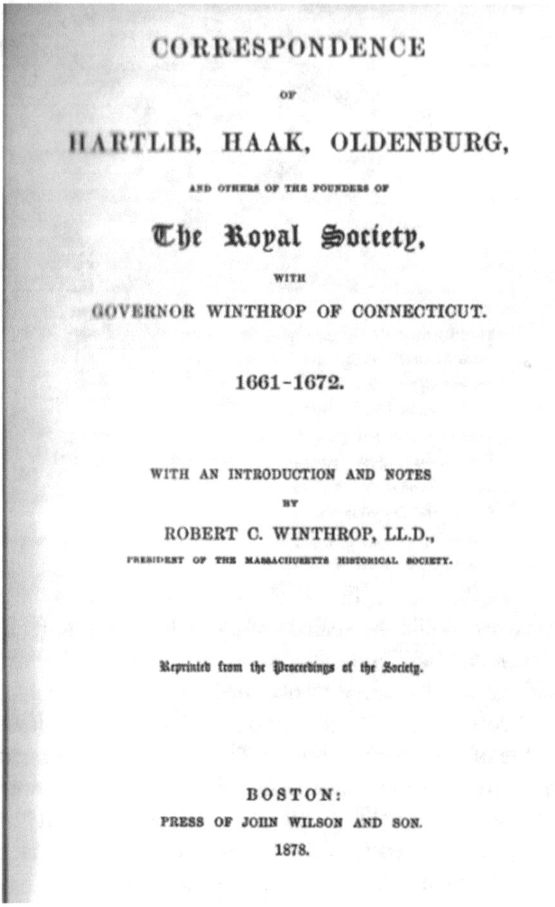

Fig. 8.2 Title page from the collection of correspondence of Governor Winthrop (scanned from personal copy)

CORRESPONDENCE

OF

HARTLIB, HAAK, OLDENBURG,

AND OTHERS OF THE FOUNDERS OF

The Royal Society,

WITH

GOVERNOR WINTHROP OF CONNECTICUT.

1661-1672.

WITH AN INTRODUCTION AND NOTES

BY

ROBERT C. WINTHROP, LL.D.,

PRESIDENT OF THE MASSACHUSETTS HISTORICAL SOCIETY.

Reprinted from the Proceedings of the Society.

BOSTON:
PRESS OF JOHN WILSON AND SON.
1878.

and "Two Astronomical Descriptions of the Comet of 1664" (the full list is given in Stearns [2]). The treasures and an accompanying letter from John Winthrop, Jr., were presented to the Royal Society on February 10, 1670. One of the New England ocean creatures was a starfish that led to a presentation before the King and extensive discussion by the Fellows. Winthrop continued to send gifts to the Royal Society and they appear in the official catalog by Nehemiah Grew (1641–1712) published in 1681 [5] (Fig. 8.3). The last known shipment was sent in 1671 in the care of his son, Waitstill Winthrop (1642–1717).

With such lavish examples of the generosity of John Winthrop, Jr., towards the Royal Society, it might be a surprise that more arcane chemistry was not a subject of his letters. Not all Fellows were alchemists, nor were they all "godly."

Clearly industrial processes were readily communicated, but occult knowledge was guarded. The name of John Winthrop, Jr., was not forgotten and in 1741, upon the occasion of a large gift from his grandson, John Winthrop (1681–1747, FRS

Fig. 8.3 Title page of
Nehemiah Grew's
catalogue (scanned from
personal copy)

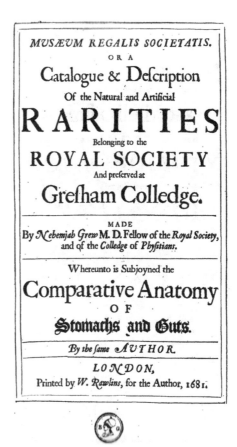

MUSÆVM REGALIS SOCIETATIS.

OR A

Catalogue & Defcription

Of the Natural and Artificial

RARITIES

Belonging to the

ROYAL SOCIETY

And preferved at

Grefham Colledge.

MADE

By *Nehemjah Grew* M. D. Fellow of the *Royal Society*,
and of the *Colledge* of *Phyfitians.*

Whereunto is Subjoyned the

Comparative Anatomy

O F

Stomachs and Guts.

By the fame AUTHOR.

LONDON,

Printed by *W. Rawlins,* for the Author, 1681.

1734), Dr. Cromwell Mortimer, Secretary of the Royal Society commented in the Dedication to Volume 40 of the *Philosophical Transactions*:

> His distant abode from *London* and his not putting his name to his writings made him not so universally known as the *Boyle's*, the *Wilkins's*, or the *Oldenburg's* of his days, nor his name handed down to us with such general applause. In concert with these and other learned friends (as he often revisited *England*) he was one of those who first formed the plan of the *Royal Society;* and had not the civil wars happily ended as they did, *Mr. Boyle, Dr. Wilkins* (as may appear in letters from *Mr. Boyle, Dr. Wilkins, Sir Kenelm Digby, &c,* to Mr. Winthrop), with several other learned men, would have left England, and out of esteem for the most excellent and valiant Governor, John Winthrop, the Younger, would have retir'd to his new-born Colony, and there established that Society for *Promoting Natural Knowledge,* which these gentlemen had formed as it were in embryo among themselves, but which afterwards, obtaining the protection of King Charles II, obtained the style of Royal, and hath since done so much honour to the British nations.

John Winthrop FRS was the guardian of many of the letters John Winthrop, Jr., had written or received from members of the Royal Society. A tabulation from the Dedication is given in Fig. 8.4.

* As might appear from the great Treafure of curious Letters on various learned Subjects ſill in your Hands, E. gr. from

Earl of Anglefey.	Dr. Everard, Ox.	Ds. Jeffe.	Prince RUPERT.
Earl of Arundel.	Pet. Jo. Faber, M.D.	Joh. Keppler.	Dr. Sackvile.
Elias Afhmole, Efq;	Monfpelii.	J. S. Kuffeler, M.D.	Earl of Sandwich.
Rob. Boyle, Efq;.	Gal. Galileo.	Dr. Lovell, Ox.	Dr. G. Starkie.
Tycho Brahe Otto-	J. Rud. Glauber.	Earl of Manchefter,	Lord Say and Seal.
nides.	Dr. Goddard.	P. Lord Chamberlain.	J. Slegelius in Ac.
Lord Brereton.	PRINCEPS Gothar.	Dr. Merret.	Franc. Med. &c.
Lord Brookes.	Dr. Grew.	Dom.Michael, Mona-	Prof.
Lord Brounker.	Mr. Hartlib.	chus.	Sir Rob. Southwell.
Dr. Browne.	Dr. Haversfeild.	John Milton.	Dr. Sprat, Bifhop of
Jo. Camden.	J. Bapt. van Hel-	Joh. de Monte.	Rochefter.
Dr. Charlton.	mont, cui fuit unius	Sir Rob. Moray.	PRINCEPS Sultsber-
Dom. Chartes Jefuita	vefperi amicus ille	Lord Napier.	genfis.
Ludg. Bat.	mirificus.	Mr. Ifaac, afterwards	Ds. Tanckmarus,
Lord Chan.Clarendon	J. Fred. Helvetius.	Sir Ifaac Newton.	M. D.
Dan. Colwall.	Lord Herbert.	Mr. Oldenburgh.	Jo. Tradefcant.
Ds. Comenius.	Hans Albrecht Do-	Dr. Pell.	Sir Philiberto Ver-
King CHARLES II.	minus Herberftein	Earl of Pembroke.	natti.
O. CROMWELL.	& Præfectus Val-	Pet. Peregrinus Romæ.	Dr. Wilkins, after-
Arthur Dee, M. D.	matiæ.	Albert Peterfon Am-	wards Bifhop of
Jo. Dee, Jun.	Joh. Hevelius Cos.	ftel. vixit an. 190.	Chefter.
Dr. Dekinion.	Gedan.	Sir Edward Peto.	Dr. Willis.
Sir K.minm Digby.	Sir Jo. Heydon.	Ds. Polemannus.	Dr. Witherly,Pr.Col.
Corn. Drebelius.	FRED. Princeps Hol-	Ju. Ray.	Med. Lond.
ERNESTUS Coloniæ	fatiæ & D. Slefvic.	Conrad. Roves Do-	Dr. Worfeley.
Epifcopus &	Robert Hooke.	minus Rofenftein	Sir Henry Wotton.
ELECT.	Ch. Howard, after-	Margrav. in Croa-	Sir Chrift. Wren.
Joh. Efpagnet, Pr.	wards Duke of Nor-	tia.	
Parl. Aquitan.	folk.		

Many of which you have given me the Pleafure of perufing; befides a great Number which it would take up too much Room here to recite.

Fig. 8.4 List of correspondents of John Winthrop, the Younger, appearing in *Philosophical Transactions* XL (scanned from original)

In commending John Winthrop FRS, Mortimer wrote in the same Dedication:

> You, Sir, have imitated the Example of your worthy Ancestor: Your Regard to the *Royal Society* shew'd itself from your Youth; you having sent to *England* many rare Curiosities for the *Museum* of the *Royal Society*, which, although by the Disingenuity of the Pilot they miss'd their Port, and were not laid up in the intended *Repository*, are some of them to be seen in a recent *Museum* now at *Cambridge*.

> … for soon after your being chosen a *Fellow*, you increased the Riches of their *Repository* with above Six hundred curious Specimens, chiefly in the Mineral Kingdom, accompanied with a List containing an accurate Account of each Particular; thereby shewing your great Skill in natural Philosophy, and at the same time intimating to *England* the vast Riches which lie hidden in the Lap of her principal Daughter.

> The extraordinary Knowledge, you have in the deep Mysteries of the most secret *Hermetic Science*, will always make you esteemed and courted by learned and good Men.

One of the most significant facts revealed by this volume of the *Philosophical Transactions* is the continuing admiration for alchemy and its adepts. John Winthrop

Fig. 8.5   Image of John Winthrop (1681–1747) (Massachusetts Historical Society, by permission)

FRS had indeed continued in the tradition of his father, Waitstill Winthrop, and his grandfather. A contemporaneous image of John Winthrop FRS is shown in Fig. 8.5.

8.1   Cotton Mather and the Royal Society

After the death of John Winthrop, Jr., life in New England became much more difficult. There was internal dissension, Native American warfare, and a desire of the English crown to "reduce New England." Eventually a Royal Governor tried to enforce allegiance to the crown and submission to English governance. However, within this chaos, there were places where learning thrived. One of the most notable figures in American history was Cotton Mather (1663–1728, FRS). He entered Harvard at 13 and could already read and write Latin, Greek and other foreign languages. While his primary activity was as a Puritan minister, he devoted himself to the life of the mind and soul, including natural philosophy and medicine. His *Christian*

Philosopher (1721) makes difficult reading today, but it reflects a deep understanding of natural philosophy. During the controversy over vaccination (when will it ever end), Cotton Mather promoted its use to save people, even with a small risk for some. He was a practicing alchemist.

His connection with the Royal Society of London developed out of his love of *Curiosa Americana*. He collected many examples of American novelties and sent them, along with an extensive discussion, to the Royal Society in the early 18th century (from 1712 to 1729). His voluminous correspondence led to his election as a Fellow, although he never traveled to London to be formally inducted.

Fig. 8.6 Title page of the *Magnalia Christi Americana* (scanned from personal copy)

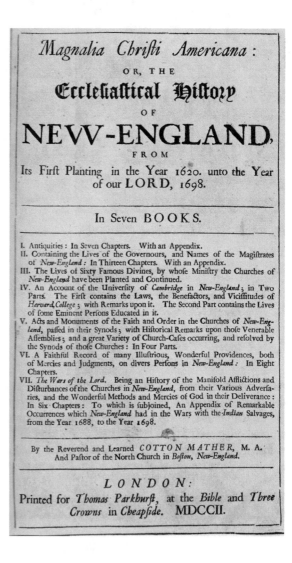

His *magnum opus, Magnalia Christi Americana* (1702) (Fig. 8.6) is one of the strangest books written in America, both then and now. Its style is indescribable, but contains a plethora of allusions in many languages and is florid beyond imagination. Nevertheless, it is an invaluable resource for the historian. It is not always accurate, but it is a unique document of the state of New England scholarship at the end of the 17th century.

In Book II, Chapter XI, there is a celebration of John Winthrop, Jr. It is entitled: *"Hermes Christianus.* The LIFE of JOHN WINTHROP, *Esq.*; *Governor of* CON-NECTICUT *and* NEW-HAVEN *United."* The orthography of Mather's works defies logic. His praise of the medical efforts of John Winthrop, Jr., is typical: "Such an one was our **Winthrop**, whose Genius and Faculty for *Experimental Philosophy*, was advanced in his *Travels* abroad, by his Acquaintance with many Learned *Virtuosi.* One Effect of this Disposition in him, was his being furnished with *Noble Medicines*, which he most Charitably and Generously gave away upon all Occasions; insomuch that where-ever he came, still the Diseased flocked about him, as if the Healing Angel of *Bethesda* had appeared in the place; and so many were the *Cures* which he wrought, and the *Lives* that he saved, that ....".

While Cotton Mather is remembered today primarily for his unfortunate connection to the Salem witch trials, he should be celebrated more for his devotion to natural philosophy and the betterment of New England society. He was a practicing alchemist and iatrochemical physician who encouraged many other Harvard students to pursue this path.

References

1. Black RC (1966) The Younger John Winthrop. Columbia University Press, New York
2. Stearns RP (1970) Science in the British Colonies of America. University of Illinois Press, London
3. Evelyn J (1664) Sylva, or, a discourse of forest trees and the propagation of timber in His Majesties dominions. Royal Society, London
4. Winthrop RC (1878) Correspondence of Hartlib, Haak, Oldenburg, and others of the founders of the Royal Society with Governor Winthrop of Connecticut, 1661–1672. Massachusetts Historical Society, Boston
5. Grew N (1681) Museum Regalis Societatis, or a catalog & description of the natural and artificial rarities belonging to the Royal Society and preserved at Gresham College. Royal Society, London

Chapter 9
The Winthrop Circle in New England

Abstract John Winthrop, Jr., was a prodigious correspondent. In addition to his English and European correspondence, many volumes of his American letters have been collected by the Massachusetts Historical Society. A discussion of his primary correspondents is presented: Thomas Shepherd, Robert Child, George Stirk, Richard Leader, Jonathan Brewster, Gershom Bulkeley, and Emmanuel Downing.

John Winthrop, the Younger, was a member of two notable correspondence circles in Europe: (1) the Hartlib Circle active from 1635 to 1659 and (2) the Royal Society circle of Henry Oldenburg. These two "intelligencers" served to connect all the active members of these communities through mutual correspondence. It is no surprise that John Winthrop, Jr., served this role in Colonial New England. His correspondence has been collected and published and serves as a vivid picture of the life of an intellectually vigorous member of the New England world [1].

John Winthrop, Jr., did not publish any complete books detailing his mature thoughts on any subject. He did publish a few papers in the Philosophical Transactions of the Royal Society, based on either papers presented directly before the Society or submitted by correspondence. This fact is consistent with his basic approach to Natural Philosophy as a privileged community of dedicated "adepts." Such knowledge needed to be protected from crass commercial interests. The spirit of the Hartlib Circle and the Royal Society stressed public revelation of the truths of Natural Philosophy. This placed Winthrop in tension with his two correspondence circles, but he readily communicated his results and his thoughts with members of these communities.

When John Winthrop, the Younger, had firmly established his Plantation at New London, Connecticut, he became the central figure in Natural Philosophy in New England and the locus of world correspondence. Upon his death, no immediate figure emerged to carry on this role, and progress in Natural Philosophy languished for a generation. The light of his life did continue to inspire a small group of fellow citizens of New England. The primary evidence for the existence of the Winthrop Circle is contained in his voluminous correspondence. This chapter will detail many Colonials that were considered chemically and spiritually acceptable as adepts.

G. Patterson, *Chemistry in 17th-Century New England*,
SpringerBriefs in History of Chemistry,
https://doi.org/10.1007/978-3-030-43261-4_9

9.1   Thomas Shepard (1605–1649)

Thomas Shepard met John Winthrop, Jr. on the ship from England to Boston in 1635 (see Chap. 5). He was a dispossessed Puritan minister fleeing from Bishop Laud. Soon after landing in Massachusetts, he was chosen as the pastor of the reconstituted Congregational Church in Newtowne (Cambridge). He was already well-known to many of the Puritan ministers in New England and had served in the same church in England as Thomas Welde and Thomas Hooker. He became one of the most beloved ministers in Massachusetts history, and is memorialized in *Builders of the Bay Colony* (1930) [2].

He was highly educated and had obtained a Cambridge A. M. in 1627 from Emmanuel College, Cambridge. While he devoted himself to his clerical duties and was one of the best preachers of the Colonial period, he never lost his love of learning. As a Congregational minister in Colonial New England, he was responsible for the entire life of his congregation. Fortunately, he was known to the Winthrop family from his time in Essex, and John Winthrop, Jr., helped him to become a competent alchemical physician.

From his office as the minister in Cambridge, Massachusetts, he was appointed to the Board of Trustees of the newly formed Harvard College [3]. The academic standards at Harvard were designed to be comparable with Emmanuel College. In addition, Harvard students listened to him preach three times each week. The classical curriculum included the *trivium* (grammar, rhetoric, and logic) and the *quadrivium* (arithmetic, geometry, astronomy and music), and the intent was to produce educated clergymen for New England churches. The natural philosophy presented in the course on physics inspired the study of the science of matter: alchemy [4].

Although Harvard got off to a rough start, the appointment of Henry Dunster (1609–1659) as President in 1640 helped to restore the college to a valuable institution. Thomas Shepard played an important part in the full life of Harvard. He helped to establish the curriculum with President Dunster and supported the teaching in his preaching. He provided sound advice to both tutors and students. And he encouraged the students to think about all of reality. Puritan clergy needed to be the best educated men in New England, so that they could accurately discern God's hand in nature and history. The Puritan elect were candidates for the revelations prepared for adepts. While modern histories of early Harvard focus on the goal of preparing theological students, the actual college tried to prepare true scholars of all knowledge. The clergymen also often became physicians [5], and the other Harvard students were acknowledged in England as having degrees on a level with Oxford and Cambridge. Many served in the Parliamentarian government and later in the court of Charles II.

The area of greatest cooperation between Thomas Shepard and John Winthrop, Jr., was in the evangelization of the Native Americans. Shepard noted that they trusted their *Pawwows* (witch doctors, shamans) for their health care. John Winthrop, Jr., made it possible to provide better health care and Shepard took advantage of this modality in his vigorous efforts to convert the Indians [3] (Fig. 9.1).

Fig. 9.1 Title page of *The Clear Sunshine of the Gospel* (1648) (scanned from personal copy)

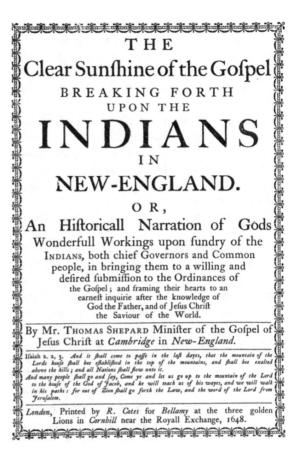

9.2   Robert Child (1613–1654)

Robert Child met John Winthrop, the Younger, in 1638 during his three year sojourn in New England. They became close friends and correspondents. Child returned to New England in 1645 and became involved in Winthrop's iron and black lead projects. Unfortunately, Child chose to challenge the political powers and escaped back to England in 1647. He made a lasting impression and is included in the classic work by Samuel Eliot Morison: *Builders of the Bay Colony* (1930) [2].

Robert Child was highly educated and graduated A. M. from Corpus Christi College, Cambridge, in 1635. He followed this with a residence at the University of Leyden in medicine, but he obtained his M.D. from Padua in 1638. In addition to his medical studies, he traveled widely in Europe and met with alchemists. By the time he met John Winthrop, Jr., he was a practiced adept and they carried out experiments together. Child was also highly interested in agriculture, and he shared this with Winthrop as well.

One of the subjects they shared was a love of alchemical books. Child obtained many of them and shipped them to Winthrop in New England. Child included a Catalogue of his own "Chymicall books." A detailed article discussing this list is: Wilson, W. J. (1943) J. Chem. Ed., XX: 123–129. One hundred twelve books and manuscripts are listed and include titles in German, Italian, French, English and Latin.

During his second stay in New England, he established a working alchemical laboratory and greatly influenced the recent Harvard graduate, George Stirk [6]. He shared many alchemical writings with this eager adept. In 1650, Stirk (Starkey) followed Child back to England. He brought enigmatic manuscripts that he claimed had been "given to him" by an adept in New England. While we now know that Starkey wrote them himself, under the pseudonym Eirenaeus Philalethes, it was long thought that either Robert Child or John Winthrop, Jr., were the authors of these works.

Quite a number of letters between Robert Child and John Winthrop, Jr., have been preserved and appear in the official Winthrop Papers published by the Massachusetts Historical Society [7]. In 1641, Child is back in Europe and is gathering materials to bring back to New England. He discusses alchemical books, grape vines from Burgundy, the progress of the Parliament, and the general situation in Europe. He hopes to collaborate with Winthrop on mining.

During John Winthrop, Jr.'s trip to England to capitalize his proposed ironworks, Robert Child became an investor. In June of 1643 he sent a letter confirming his joy in Winthrop's safe arrival back in New England. Child purchased 5 tun of furnace stone and sent them to Boston. He is worried that Winthrop's late arrival in the year will postpone any serious work on the ironworks. He inquires about the black lead mine. He was very active in procuring the needed parts for the ironworks. He announces that he will come to New England himself in the Spring. Nevertheless, the next letter in March 1645 finds Child still in Gravesend [1]. Child and Winthrop engaged in a seed exchange for the benefit of both England and New England. Child also sent vines and shrubs like pyrocanthus. He still hopes to come to New England in Spring 1645. He continues to be active in England for the ironworks and has been in contact with Richard Leader (see Chap. 6). He promises to help fund the black lead efforts. He also gives a detailed alchemical discussion of bismuth, lead, tin and mercury and their mining. He references Libavius as an authority. He also discusses black lead as an earth like chalk and its infusibility. Robert Child was a serious chemist and was in contact with many of the adepts in England and Europe. John Winthrop, Jr., was both his good friend and fellow adept.

In March of 1647, Robert Child is in Boston and has been very active in the ironworks at Saugus and at "Brayntree." John Winthrop, Jr., is in Pequot at his new plantation. But, there is a "problem." John Winthrop, Sr., has imprisoned Child under "house arrest." He acknowledges that it is because of the "Remonstrance." He is in need of money and requests the return of a loan. If he does satisfy the Massachusetts authorities, he proposes to travel to Long Island and start a glass business in Dutch New Amsterdam. He remains on affectionate terms with Winthrop, but is clearly concerned for his future.

A letter of May 13, 1648 finds Child back in England at Gravesend. He discusses the production of "Turpentin" from French pitch pines. He notes with sadness that since the New England authorities have robbed him of a great fortune, he cannot invest any more money there. He is still buying alchemical books both for himself and for Winthrop, including works by Glauber and Van Helmont. He notes that he is friends with Sir William Petty (1623–1687, FRS), who had just invented a pantograph. (Both Child and Petty pursued a fortune in Ireland during the Parliamentarian period and obtained it!) He mentions that a friend in Scotland (unnamed) has perfected the "menstruum of Helmont" and has made progress on transmutation. Child encourages Winthrop to continue the seed and mineral exchanges. Child also claims that he has heard that "Helia Artista" has been borne (the Messiah of the Rosicrucians).

In another letter (perhaps delayed) Child notes his place of residence for the summer of 1648 will be with Dr. Garbet at Hogs Downe. He plans to carry out alchemical experiments with his host. Then he plans to settle in Kent. The Civil War in England has reached a very bad point, with many innocent people being killed. There is war all over Europe in those days.

In March 1649, a letter from Winthrop to Child recounts his life at Pequot. He discusses his desire to obtain the books listed by Child in his letter of 1648. (If Child had lived, he would certainly have been a founding member of the Royal Society. In the event, he died in 1654.)

The last letter in the compilation is from Child to Winthrop on August 26, 1650 from Gravesend [8]. He discourses on the relation between Van Helmont and Paracelsus, and frankly prefers the more ancient authority. He highly values Glauber, but has not yet obtained any of his books. Child was a physician and notes a new book by William Harvey. He is still is communication with Richard Leader about alchemical books as well. One of the most interesting subjects in this letter is the proposal to "retreat to a more solitary life, as I can Command myself, with 6 or 7 gentlemen and scollars, who have resolved to live retyredly and follow their studies and Experiences, if these troublesome times molest not. These gentlemen for Curiosity and Learning scarcely have their equals in England."

Robert Child's life in England was rich and interesting after his return. He associated with the Hartlib Circle, and became close friends with John French and John Milton. He wrote a three volume treatise on "Defects and Remedies of English Husbandry." He became heavily involved in the agricultural development of Ireland in the Parliamentarian period and was in close touch with Robert Boyle, who also had Irish estates. Child lived with Colonel Arthur Hill at Hillsborough Castle near Belfast, and was working as an agricultural consultant. Hartlib had high hopes that Robert Child would complete a Natural History of Ireland, but he died in 1654.

9.3 George Stirk (1628–1665)

George Starkey met John Winthrop, Jr., while he was a student at Harvard, from which he graduated in 1646. He obtained an A.M. in 1649. During his postgraduate years he pursued alchemy with help from Child and Winthrop.

The life of George Starkey is brilliantly presented by William R. Newman in *Gehennical Fire: The Lives of George Starkey* (1994) [6]. He was the son of a Bermuda clergyman, George Stirk (1595–1637). His father taught him the ancient languages and the son explored the fauna of Bermuda with more than juvenile skill. Upon his father's death, another Bermuda minister, Patrick Copeland, wrote to Governor John Winthrop, enquiring about a college for young George. He matriculated at Harvard in 1643 and received his A.B. in 1646. He was made A.M. in 1649.

The curriculum established by President Henry Dunster (1609–1659) was modeled after his own Cambridge experience at Magdalene College (A.M. 1634). It included much more than the basic languages and classical subjects. Of special note is the course on Physicks [4]. In the seventeenth century, this was the full subject of natural philosophy, including matter theory. In addition, George Stirk's tutor was President Dunster himself. (I would have loved to have been present for these weekly sessions.)

The New England community was tightly knit and families were often united by marital ties. One of the early tutors of Harvard was George Downing (1625–1684, M.A. 1645), John Winthrop, the Younger's cousin. While Winthrop was very busy during the years 1643–1646, some of his library and many of his chemicals were in Boston at his father's house. When a Charlestown physician, Richard Palgrave (d. 1651), encouraged Starkey in 1644 to study alchemy, he would have had access to the Winthrop legacy. Palgrave was one of the original passengers on the *Arbela* in 1630, and knew Governor Winthrop well. By 1644 he would have interacted with John Winthrop, Jr., extensively.

While Starkey's initial efforts in the world of alchemy were based on arcane books and academic disputations, he soon realized that in order to understand this subject, one needed to become a devotee of the furnace. He built his own furnace and pursued experimental alchemy with Palgrave and another of his Harvard mates, John Alcocke (1627–1667) (Class of 1646, M.A. 1649). Later, Starkey sought materials and glassware from John Winthrop, Jr. in a letter of 1648 [1].

"I hear you shortly intend to come to the bay, if by water, if you could spare any antimony and mercury, I should content you for it and rest engaged. If you could spare one or two of your greater glasses you would doe me great pleasure. I wish if you could find Helmont *de febribus*, I might borrow him of you as also *de lithiasi*, Also the little book intituled Encheiridion Philosophaie restitute with Arcanum Philos; at the end of it. If your Worship would be pleased to remember the keys of the cabinets wherein your books are, I should count it an extreme felicity once to have the view of Chemical books which I have not read a long time. Theatrum Chemicum I should chiefly desire. I have built a furnace, very exquisitely but want glasses, and antimony and mercury." Aug. 2, 1648.

Upon receiving his Bachelor's degree, Starkey established a medical practice in Boston. This successful venture provided funds to pursue his real passion: alchemy. Palgrave and Winthrop were alchemical physicians and Starkey realized that there were very few practicing physicians in New England. Many of the classic alchemical books discussed the use of chemicals in the practice of medicine: *iatrochemistry*.

George Starkey emigrated to England in 1650 and became one of the most famous alchemists. One of his closest friends in New England was Robert Child. They both discussed and practiced alchemy. But, Child was driven out of New England in 1647 after remonstrating with the Colonial authorities about their religious and political intolerance. It is not at all clear that Child and Starkey collaborated in England, even though they were both friends of Robert Boyle. And Child died in 1654. Starkey died in 1665, while treating plague patients in London. (The Royal Physicians had fled! Starkey would not have been allowed to practice medicine in London under ordinary circumstances.)

9.4 Richard Leader (1609–1661)

Although Richard Leader originally met John Winthrop, Jr., under circumstances that could have been strained, they became good friends and frequent correspondents. In 1645, he was recruited to take control of the Ironworks in New England that was started by Winthrop. He founded a new site at Saugus and brought the Hammersmith works to completion [9] (see Chap. 6).

One of Leader's notable roles in American history was his hospitality to Robert Child during his "house arrest" during 1646–8. Child was thrilled to discover Leader's magnificent library consisting of many "curious" books on chemistry and religion. Child was an investor in the Company of Undertakers for the Ironworks in New England and helped with the operation during his confinement. He also carried out analyses of ores and pursued alchemy with Leader.

In a letter of August 21, 1648 from Hammersmith to "his very Loving friend John Winthrop Junr. Esqr" Leader sends greeting from Child and begs some mercury for analysis. He also complains that relations with the Undertakers in England have soured and he does not anticipate a good outcome [1].

Eventually, Richard Leader was replaced as the manager of the Hammersmith Forge. But, Leader was very enterprising, and moved on to build a sawmill in Kittery, Maine. Like his friend, Robert Child, Leader fell out with the Massachusetts authorities and was forced to pay a heavy fine. While Child was a Presbyterian during the Parliamentarian era, Leader was a much more arcane follower of two English "Prophets," John Reeve and Lodowick Muggleton.

Richard Leader linked up again with John Winthrop, Jr., in a saltmaking enterprise in Barbados (1659). While the technical side of this enterprise was very sophisticated, the hurricanes that plague this region put an end to the saltworks.

In his final letter to John Winthrop, Jr., in 1661, Leader reflects on the state of England and of the future of New England, where he died.

"I question not but you are fully informed of the great change and sudden alteration in England. If the tender hearted ones are deprived the liberty of their consciences, to serve their God in truth of heart, but must all be forced to fall downe to what shall be Established by a State, my face shall be turned away from ever having thought to see my native country while I live [9]".

While Richard Leader and John Winthrop, Jr., made an odd couple, they shared many deep convictions and joined in many industrial and alchemical adventures.

9.5   Jonathan Brewster (1593–1659)

Jonathan Brewster was the son of the famous Pilgrim father William Brewster of Plymouth (1566–1644) [10]. After his father's death, he established himself in Connecticut as a trader. He joined John Winthrop, Jr., at Pequot (New London) in 1649 [1]. Jonathan Brewster is mentioned in an early list of honored residents of New London [11]. He was elected as the Town Clerk in 1650 and was made a Freeman of the Colony of Connecticut.

Jonathan Brewster was an "Indian Trader." He established a trading house on the Pequot River at a place still called "Brewster's Neck," on the Eastern side of the river, opposite to Mohegan. The deed to this property was issued by the "Sachem of Mauhekon," Uncas. Eventually, this large tract (700 acres) was incorporated within the town of Norwich. In the finest baronial tradition, Brewster was granted a monopoly for the Indian trade. Jonathan Brewster proved to be a good friend to Uncas, and ably defended him during a siege by the ferocious Narragansett Indians.

Jonathan Brewster carried on an active correspondence with John Winthrop, Jr., and often shared both information and chemicals. The most remarkable letter in this series was sent on January 31, 1647 [12]. It is discussed in detail in *Prospero's America* [13]. Brewster was a serious alchemist and often spent himself into debt. But, he was proud of his accomplishments. He was also worried that he might die (not an unreasonable fear in seventeenth century Connecticut). He wanted to pass on his knowledge to a trusted adept so that it would not be lost. But, he did not want to publish it to the world at large. He trusted John Winthrop, Jr.

He claimed to have discovered "how to work the Elixer, fit for Medicine, and healing of all maladyes [13]." He was confident that "my worke being trew thus far, by all their writings, it cannot faylle." He had been working on the project for four years, and anticipated it would take three more to come to completion. (Alchemy was not for the impatient!) Although he was the son of a serious Puritan, he did not stay in Plymouth or settle in New Haven. He was completely devoted to the spirit of alchemy.

9.6 Gershom Bulkeley (1636–1713)

Gershom Bulkeley was the son of the Reverend Peter Bulkeley, one of the displaced Puritan ministers who sailed in 1635 to New England with John Winthrop, Jr. [14]. He attended Harvard University (Class of 1655) and prepared to become a Puritan minister. He had the good fortune to be established at New London in 1661. He served the local people well, but by 1666 he moved to Wethersfield, CT [11]. While he was in New London he learned a new skill, which he practiced for the rest of his life. He became an alchemical physician, in the model of John Winthrop, Jr.

Of the many roles that Gershom Bulkeley played in the life of seventeenth century New England, the most important was as the physician to the militias. He also served as the chaplain (two for the price of one!). He resigned from his Wethersfield pulpit in 1677 and practiced medicine and politics for the rest of his life. Upon his death he was eulogized by the Boston News: "Eminent for his great Parts, both Natural and Acquired, being Universally acknowledged, besides his good Religion and Vertue, to be a Person of Great Penetration, and a sound Judgment, as well in Divinity as Politicks and Physick; having Served his Country many Years successively as a Minister, a Judge, and a Physitian with great Honour to himself and advantage to others [14]." But, perhaps our most useful memorial to Gershom Bulkeley was an elegiac poem dedicated to him after his death [14] (Fig. 9.2).

9.7 Emmanuel Downing (1585–1660)

Emmanuel Downing was uncle to John Winthrop, Jr. He had married his father's sister Lucy. They served as guardians while Winthrop was at Trinity College, Dublin. Emmanuel was a lawyer at the Middle Temple in London. He exchanged dozens of letters with his nephew during his lifetime and faithfully served the family in many roles. In return, John Winthrop, Jr., shared his knowledge, expertise and political power with his uncle. Emmanuel invested in the Ironworks and negotiated for his nephew while in England. He settled in Salem, MA and carried out a variety of chemical enterprises, especially the distillation of flavorful strong waters! He kept his eyes open for any scheme that might actually make money. One of the most interesting letters contained a recipe for making indigo dye [8] (Fig. 9.3).

Fig. 9.2 Elegiac poem honoring Gershom Bulkeley attributed to Johannes Jamesius Londoninensis [14]

"On the DEATH of the very learned, Pious and Excelling | Gershom Bulkley Esq. M. D. | Who had his Mortality swallowed up of Life, *December* the *Second* 1713. Aetatis Suæ 78. |

"Sanctus erat Quanquam Lucas, Medicusque, Sepulchri
Jura subit, factus Victima dira necis:

A Saint tho' Luke, and a Physician too,
Struck Sail to Death, as other Mortals do.

[1]

"HOW vast acquests of Learnings store ⎫
Had he amass'd! still gathering more: ⎬
Resolv'd therein ne'r to be Poor. ⎭
Jurist, Divine, and Med'cines Votary ⎫
Where's he in each him matcht, or came but nigh ⎬
That had them all in a Transcendency? ⎭
His Graces and his Vertues brave
A Golden tincture thereto gave:
And do perfume his Precious Name,
That all who know and hear the same;
Thereto such Epithets will give,
That he tho' Dead, Renown'd will Live.

[2]

"*Gershom*¹ no more! Fatigues & Hazards past:
He's safe arriv'd to th' Promis'd Land at last.
In Heavens Academy, he
Adeptist: O how glad to be!
Where none do longer rack their Brains
In quest of Scientifick Gains.
He in a Nobler Orb does move
Encyclopedian Tract Above;

¹ Exodus ii. 22.

That Atmosphere beyond now got ⎫
(Farewell bid to *Connecticot* ⎬
Of Revolutions strange the spot) ⎭
Has in Immanuel's Land his Lot: ⎭
Where the dire and malignant Aspects fail
Shed from Medusa's Head and Dragons Tail."

"A Pure Extract and Quintessential wrought,
The *Caput Mortuum* is hereto brought.
Brave Chymist Death! how Noble is thine Art?
The Spirits thus who from the Lees canst part,
'By Sacred Chymistry the Spirit must
'Ascend, and leave the Sediment to Dust.'"

ENCLOSURE:
INDIGO. THE RECEIPT FOR MAKING OF INDIGO

1 or 2 houres after the herb is cutt lay it in a fatt presse it downe hard with a beame over cross barres that aire may come to it till it worke and raise the barrs. Let it lye 24 houres then fill the fatt halfe full of water till the weede rott in the water usually in 24 houres, then fill the fatt full. So lett it stand untill it come to a coulor within 3 daies tyme, the weede unrotted take out. Lett the rest stand 24 houres more then stirre it, that it may all runne out into an other fatt: there beate it and poure it in and out with bucketts and that incessantly, till it come to one perfect coulor; lett it then settle, make then a tap to draw forth all the thin water, then take up the bottom remaining into baggs that will hold 5 pound weight, made of strong canvasse with an hoope on the top, and then a stick a crosse, by which hang it in an house and save the droppings, which will make a good couler (so the first drawne water a reasonable coulor) in an houres tyme the water will all dropp out of the bagge, then take the remaining Indico into boxes, in which lay the Indico some 3 fingers thick, which set in the sunne, and lett them candy[1] (else in an oven or stove to dry temperately not in hast),[2] then whilest it is drying slice it with a knife.[3]

Memord: the vine cotton like to grow heere.

FC; docketed by JW2 "of Indigo." FC in JW2's hand throughout. Missing original enclosed in Downing to JW2, March 13, 1653/54, directly above.

[1] To form into crystals or congeal in crystalline form; see *OED*, candy (as verb), 3.
[2] Closing parenthesis supplied.
[3] Indigo was a dyestuff early regarded as "of extraordinary value" to England (*Cal. State Papers, Col., 1574–1660*, p. 168); its export from her overseas plantations was later confined to the Mother Country (Andrews, *Colonial Period*, IV, 86). Its New World growth and production were best accomplished in tropical and subtropical climates; on such in mainland North America, in a process that was "laborious, messy, and stinking," see Wright, *S. Car.*, p. 79, and Weir, *Colonial S. Car.*, p. 151.

Fig. 9.3 Recipe for making indigo sent by Emmanuel Downing [8]

References

1. Winthrop A (1947) Winthrop papers, vol V, 1645–1649. The Massachusetts Historical Society, Boston
2. Morison SE (1930) Builders of the Bay Colony. Houghton Mifflin Co., Boston
3. Shepard T (1853) The works of Thomas Shepard: First Pastor of the First Church, Cambridge, MA, with a memoir of his life and character. Doctrinal Tract and Book Society, Boston
4. Morison SE (1936) Harvard College in the seventeenth century, part I. Harvard University Press, Boston
5. Watson PA (1991) The Angelical Conjunction: the preacher-physicians of colonial New England. University of Tennessee Press, Knoxville
6. Newman WR (1994) Gehennical fire: the lives of George Starkey, an American Alchemist in the scientific revolution. The University of Chicago Press, Chicago
7. Winthrop S (1944) Winthrop papers, vol IV, 1638–1644. The Merrymount Press, Boston

8. Freiberg M (1992) Winthrop papers, vol VI, 1650–1654. Massachusetts Historical Society, Boston
9. Hartley EN (1957) Ironworks on the Saugus: the Lynn and Braintree Ventures of the Company of Undertakers of the Ironworks in New England. University of Oklahoma Press, Norman
10. Bradford W (1920) Bradford's history of the Plymouth Settlement, 1608–1650, Rendered into Modern English by Harold Paget. E.P. Dutton and Company, New York
11. Caulkins FM (1895) History of New London, Connecticut. H.D. Utley, New London
12. Brewster J (1865) Letter to John Winthrop, Jr. Massachusetts Historical Society Collections, VII, 77–81
13. Woodward WW (2010) Prospero's America: John Winthrop, Jr., Alchemy, and the Creation of New England Culture, 1606–1676. University of North Carolina Press, Chapel Hill
14. Sibley JL (1873) Biographical sketches of graduates of Harvard University in Cambridge, Massachusetts. Charles William Sever, Cambridge

Chapter 10
Harvard College and Seventeenth Century Chemistry

Abstract Contrary to a later period in the history of Harvard University, the seventeenth century produced many practicing physicians. They also served as Puritan ministers. This chapter tells their story. The three Presidents of Harvard University, Henry Dunster, Charles Chauncy and Leonard Hoar, were all physicians. They mentored budding iatrochemists during both their undergraduate and graduate studies. In addition, the students apprenticed with practicing physicians in the Boston area such as Richard Palgrave. This chapter presents brief vignettes of Cotton Mather, Michael Wigglesworth, Edward Taylor, Samue Lee and Thomas Palmer.

Harvard College was founded to provide an elite education for the children of the New England elite. The curriculum was designed to replicate a typical college at Cambridge, England. Instruction was in Latin and the students were expected to declaim and to argue in the classical tongue. Of what use was such a classical education in the new wilderness?

The primary goal of a Harvard education was to prepare new Puritan ministers for both America and eventually for England itself. In this it succeeded. After the students had obtained the B.A. degree (initially three years, eventually four years), they typically spent another three years and received an M.A. degree. This qualified them to enter the elite world of English culture. All Harvard students could qualify for a Puritan pulpit, but not all chose this opportunity. Even for those that did become covenanted ministers, the economic realities of life in New England required additional funds to survive. How could they do this? This story is beautifully told in *The Angelical Conjunction: The Preacher-Physicians of Colonial New England* by Patricia Watson [1].

Harvard College did not offer instruction in medicine in the seventeenth century. Real medical schools were in Europe in places like Leyden and Padua. How could students in Colonial New England ever hope to obtain proficiency in medicine? The path involved being trained by a currently practicing physician and reading the standard medical texts of the seventeenth century. Most of these books were written in Latin! A Harvard education prepared bright students to train to be doctors. During their post-baccalaureate years, they learned their trade from actual physicians. Many

© The Author(s), under exclusive license to Springer Nature Switzerland AG 2020
G. Patterson, *Chemistry in 17th-Century New England*,
SpringerBriefs in History of Chemistry,
https://doi.org/10.1007/978-3-030-43261-4_10

of them defended medical theses for their M.A. One of the most famous Harvard trained physicians was George Starkey (1628–1665) (M.A. 1649).

While Henry Dunster (1609–1659), the first President of Harvard, was fully capable of teaching all the classes at Harvard, he was also a practicing physician. He obtained his M.A. from Magdalene College, Cambridge in 1634 and was Headmaster at the famous Bury St. Edmonds Grammar School. (This was the school John Winthrop, Jr., attended.) He was recruited by the Reverend Richard Mather (1596–1669) to become the Master of Harvard College in 1640. He designed the curriculum and the details are well-described in the official history of Harvard College by Samuel Eliot Morrison [2]. In addition, Dunster served as the tutor for some of the students, including George Starkey. Dunster recognized the talent of the precocious Bermudian and by his second year in college he was studying with a local physician, Richard Palgrave (d. 1651) of Charlestown. By the time Starkey graduated in the class of 1646 he was ready to earn his living as a physician in Boston. This was beneficial since he was an orphan and did not have independent means or a local family with which to live. His medical practice was a success, and he used the profits to pursue his great love: alchemy.

Palgrave was an iatrochemist who synthesized his own pharmaceuticals. He arrived in New England in 1630 in the same fleet as John Winthrop, Sr. In addition to explicitly medical knowledge, Palgrave taught Starkey the art of alchemy. Apparently, Palgrave also taught John Alcock (1627–1667, M.A. 1649). Alcock carried out alchemical experiments with Starkey, established an independent medical practice in Roxbury, Massachusetts, and married Palgrave's daughter Sarah (1621–1665) in 1648. She was also a very active physician and surgeon.

While women were forbidden at Oxford and Cambridge, they were an essential part of the healing community in New England. Even if the Puritan minister himself was not a practicing physician, his wife was often called upon to render aid to congregants. John Winthrop, Jr., provided free medicines to many congregations to be dispensed by the wife of the preacher.

When it came time to elect a new President of Harvard, the Overseers looked for another Cambridge educated scholar. Charles Chauncy (1592–1672) was already in New England in the Plymouth Colony at Scituate, Massachusetts. While the Pilgrims came earlier, and were from a different branch of the Puritan movement, both groups acknowledged Puritan divines from Cambridge. Like Henry Dunster before him, he practiced medicine to augment his salary. He shared the same educational goals as well. Chauncy served from 1654–1672. One of his greatest bequests to New England was his family. The Chauncy's became preacher-physicians, leading politicians and staunch Old Light churchmen. Two of the Chauncy sons, Elnathan (1639–1684, M.A. 1661) and Israel (1644–1703, M.A. 1661) became well-known alchemists.

During the period at the end of Charles Chauncy's life, while he lingered, a search for a new President was conducted. The greatest scholar then residing in New England was Leonard Hoar [1630–1675, Harvard A.B. (1650) Cambridge M.D. (1671)]. He was both a noted Puritan minister and a practicing physician. He was encouraged to return to New England, with a view to becoming President of Harvard, and was duly elected in 1672, upon the death of Chauncy. While he knew the requisite Arts and was well-qualified to govern Harvard College, he was unaware of the resentment that his appointment would produce. The current Fellows of Harvard, one of whom expected to be made President, stirred up the students to detest Hoar. And, to make matters worse, resigned themselves and encouraged the students to withdraw. While it took three years to drive Hoar into his grave, his future at Harvard was finished. The damage to Harvard was long-lasting and was made even more tragic by the knowledge of Hoar's plans for Harvard.

While many of the Puritans were ruined by the Restoration, Leonard Hoar was wise enough to remain neutral with regard to government, even though he was ejected from his pulpit by the 1662 Act of Uniformity. Since he was a practicing iatrochemical physician and a scholar, he was made Doctor of Physick by Cambridge University in 1671. He was well-known by the Fellows of the Royal Society and was on good personal terms with Robert Boyle. In a letter of 1672 written to the Honorable Robert Boyle after he was elected President of Harvard he outlined his plans for the improvement of the college. "A large well-sheltered garden and orchard for students addicted to planting; an ergasterium for mechanic fancies; and a laboratory chemical for those philosophers, that by their senses would culture their understandings, are in our design, for the students to spend their times of recreation in them; for readings or notions only are but husky provender." The seventeenth century was filled with plans for a true community of scholars. Harvard could have been such a place, but the forces of parochial nepotism ruined the place until the eighteenth century.

10.1   Cotton Mather (1663–1728)

Even in an environment of time-serving tutors and useless governors, a few outstanding students emerged from Harvard. The most chemical of these scholars was Cotton Mather (M.A. 1678). His father, Increase Mather (1639–1723, Harvard B.A. 1656, Trinity College Dublin M.A. 1658), had a fine library that included many alchemical works. Cotton Mather was eventually elected as a Fellow of the Royal Society in 1713 [3]. One of Mather's friends, John Winthrop (1681–1747), the grandson of John Winthrop, Jr., was elected FRS in 1734. Winthrop had returned to England where he lived for the rest of his life. Although Cotton Mather was supported by the New England church he pastored, he earned additional money by iatrochemistry.

Cotton Mather was a very prodigious author, and two of his medical works included *The Great Physician* (1700) and *The Angel of Bethesda* (1724). He is most famous in the medical world for his vigorous promotion of inoculation for smallpox. The faith that inspired him to do this was based on both his Puritan principles and his knowledge of the best medical practice in England. He also inspired Jonathan Edwards to advocate for inoculation at the end of the eighteenth century. Edwards died as a result of his personal inoculation, but he saved thousands of other citizens in New Jersey as President of Princeton.

10.2   Michael Wigglesworth (1631–1705)

One of the most celebrated of the Harvard-trained preacher-physicians was Michael Wigglesworth (Class of 1651, M.A. 1654). He continued at Harvard as a tutor. He was praised by Cotton Mather: "With a rare Faithfulness did he adorn the Station! He used all the means imaginable, to make his Pupils not only good Scholars, but also good Christians; and instill into them those things, which might render them rich Blessings unto the Churches of God. Unto his Watchful and Painful Essayes, to keep them close unto their Academical Exercises, he added, Serious Admonitions unto them about their Interior State, and (as I find in his Reserved Papers) he Employ'd his Prayers and Tears to God for them, and had such a flaming zeal, to make them worthy men, that, upon Reflection, he was afraid, Lest his cares for their Good, and his affection to them, should so drink up his very Spirit, as to steal away his Heart from God." He was famous for an extended poem called *The Day of Doom*, which was popular in New England for decades. He also wrote *God's Controversy with New England*. He was the pastor of a Congregational Church in Malden, Massachusetts, but his income was mostly derived from his practice of iatrochemical medicine. He was offered the Presidency of Harvard, in respect of his scholarship, but declined [4]. He was also remembered for his "impressive medical library."

10.3   Edward Taylor (1643–1729, Harvard M.A. 1671)

Edward Taylor was also a highly literate iatrochemist and preacher. He was born in England and educated at Cambridge. He was forced out of the university in 1662 by the Act of Uniformity since he was a Puritan. He emigrated in 1668 to New England and entered Harvard College to prepare for the American Puritan ministry [5]. He settled in Westfield, Connecticut and organized a church there. He practiced iatrochemical medicine that he had learned while at Harvard. He also had a large library, both of scientific books and literary ones. He is remembered today as a great poet [6]. An example from 1683 is Fig. 10.1.

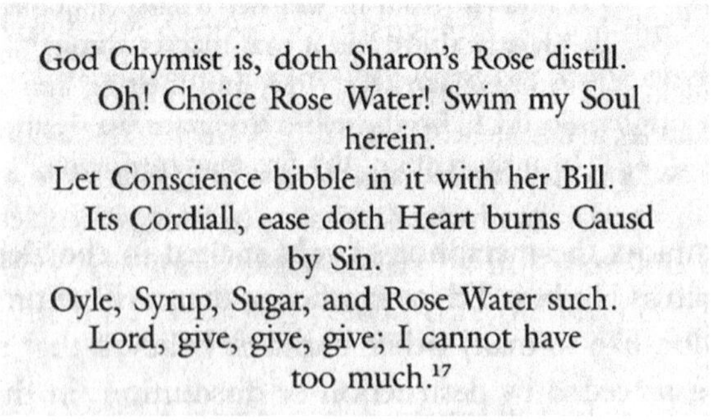

Fig. 10.1 From Donald Sanford, *The Poems of Edward Taylor* (1963)

God Chymist is, doth Sharon's Rose distill.
Oh! Choice Rose Water! Swim my Soul
herein.
Let Conscience bibble in it with her Bill.
Its Cordiall, ease doth Heart burns Causd
by Sin.
Oyle, Syrup, Sugar, and Rose Water such.
Lord, give, give, give; I cannot have
too much.[17]

Another choice stanza due to Taylor from the same collection:

Gold in its Ore, must melted be, to bring
   It midwift from its mother womb: requires
To make it shine and a rich market thing,
   A fining Pot, and Test, and melting fire.
So do I, Lord, before thy grace do shine
In mee, require, thy fire may mee refine.[20]

10.4 Samuel Lee (1625–1691, Oxford M.A. 1648)

Samuel Lee was praised by Cotton Mather as one of the most learned men to immigrate to New England. He settled in Bristol, Rhode Island and was pastor of the Congregational Church there from 1687 to 1691. He amassed a library of 1100 volumes, many of which were alchemical in nature. He practiced iatrochemical medicine and is cited by Stearns [3] as a regular correspondent of Nehemiah Grew, FRS, in England and Judge Samuel Sewell in Boston [5]. One of Samuel Lee's other colleagues in New England was William Avery, already noted as a famous physician and alchemist in Dedham and Boston, Massachusetts. The correspondence with Nehemiah Grew was largely about the practice of *physick* in New England. Unlike London, it was unregulated and there were very few members of the Royal College of Physicians practicing in New England. Most of the physicians were also their own pharmacists. They made what they could, and imported the rest from Europe.

In George Kittredge's Notes, there is an extensive discussion of William (1622–1687) and Jonathan (1653–1694) Avery [5]. Both father and son were avid alchemists. Jonathan was willed his father's extensive library and alchemical equipment.

Samuel Lee was in the middle of a return trip to England in 1691 when the ship was attacked by French pirates (a very common calamity in this time period). He was taken prisoner and died in St. Malo, France in 1692. Cotton Mather includes him among the "martyrs of the Christian Faith!" Cotton Mather's third wife, Lydia, was the daughter of Samuel Lee.

10.5 Thomas Palmer (1666–1743)

Thomas Palmer was born and raised in Situate, Plymouth Colony. He was educated locally and practiced as an alchemical physician from around 1690 until his death.

He sent two sons to Harvard and left an extensive estate. He also preached at Middleborough, Massachusetts from 1696–1708. He is most famous for the book that he wrote: *Admirable Secrets in Physick and Chyrurgery* (1696) (Fig. 10.2).

The book is a very practical manual for alchemical physicians in New England. It combines animal, vegetable and mineral remedies. It discusses diagnosis of disease, the appropriate remedies and the overall Hippocratic philosophy of medicine. While Palmer was highly successful in Plymouth Colony, he would have not been permitted to practice in England. He is not listed as a graduate of Harvard College, but he was highly respected in eighteenth century Massachusetts.

He lists the *presages* of disease that the observant physician can use to determine the illness. This careful protocol was quite sophisticated, especially in 1696. His list of remedies extends to more than 1000 substances! He was familiar with the

Fig. 10.2 Title page of *Admirable Secrets* (1696) by Thomas Palmer (scanned from personal copy)

standard "Herbals" of his time, and gathered his own specimens whenever he could. He synthesized his mineral medicines and was familiar with the Paracelsian corpus.

Situate was the home of three famous preacher-physicians: Charles Chauncy (1592–1672), Henry Dunster (1609–1659) and Leonard Hoar (1630–1675). This tradition was passed on to Thomas Palmer, even though he could not have learned it directly from them. The history of Situate notes that until 1719, the physicians were all preachers. Not only did Thomas Palmer become a highly effective physician, he left a priceless manuscript to guide others in his own time and to inform historians today.

References

1. Watson PA (1991) The Angelical Conjunction: the preacher-physicians of Colonial New England. The University of Tennessee Press, Knoxville
2. Morison SE (1936) Harvard College in the seventeenth century. Harvard University Press, Cambridge
3. Stearns RP (1970) Science in the British colonies of America. University of Illinois Press, Urbana
4. Sibley JL (1873) Biographical sketches of graduates of Harvard College in Cambridge, Massachusetts, vol 1, 1642–1658, Cambridge
5. Kittredge GL (1912) Letters of Samuel Lee and Samuel Sewell relating to New England and the Indians. John Wilson and Son, Cambridge
6. Palmer T (1696) The admirable secrets of physicks and chyrurgery. Yale University Press, New Haven (Modern edition (1984) edited by Thomas Forbes) (Manuscript)

Chapter 11
Conclusions and Reflections

Abstract A short essay on both chemistry in seventeenth century New England and the history of chemistry itself is presented. John Winthrop, Jr., relied primarily on his extensive reading and his careful empirical work to produce both needed chemicals and good recipes for others to follow. The story told in these pages is much more interesting and substantially more complicated than traditional views of Colonial America.

While standard histories of chemistry completely ignore such activities in seventeenth century New England, there was a vibrant world of chemistry on American soil. It was driven largely by the needs of a Colonial culture. The colonists had a large need for salt to preserve fish and meats. The importance of salt in the history of the human race is a continuing story [1]!

Chemistry takes what Nature gives and produces materials that humans want or need. New England was filled with trees. The chemistry of trees is still an important part of human culture, but it was all important in the seventeenth century [2]. While the early Colonists focused on the use of trees to create lumber for houses and ships, such activities also required other forest products. Trees were a source of turpentine and tar. John Winthrop, Jr., knew how to extract these chemicals from the forests of New England (with help from the local inhabitants). Trees also provided the raw materials for potash (potassium carbonate), and hence for saltpeter [3].

Trees also serve as the raw material for the synthesis of charcoal [4]. The importance of charcoal for world culture cannot be ignored. While trees were plentiful in Colonial New England, real "charcoalers" were scarce. John Winthrop, Jr., hired English charcoal workers and they trained the locals. Charcoal is still one of the most important chemical substances in twenty-first century America.

New England is blessed with minerals. Chemists take the raw minerals and process them into useful substances, such as pig iron. There were many serious problems associated with the minerals industry in New England. Perhaps the largest issue was transportation. Ideally, ore could be transported by boat. But, if the mine site was very far upriver, it would be difficult to navigate all the way to the Atlantic Ocean, or at least to the site of a smelter. There was also the issue of permission to operate a mine. The Native Americans demanded payment for the right to work in their territory. The

Massachusetts government also demanded the right to license any activity that made money. In return, they usually granted a 21 year monopoly. Also, the local town often granted both land and water rights to help run the mill. Many attempts to found a thriving minerals industry in New England were made, but, in the end, financial and management issues hindered full realization. Nevertheless, iron was produced in New England in the seventeenth century [5]!

Any substance containing sugar can be fermented into a beverage. The Puritans ran on beer, and made sure to bring good brewers to New England. There they found that corn could also be fermented, and John Winthrop, Jr., served corn beer to the Royal Society. Fermented beverages can be distilled to produce "strong drink," and John Winthrop, Jr., taught his aunt and uncle Downing how to safely and deliciously produce flavored alcohol beverages. I wonder what their corn liquor tasted like! Eventually, the most important chemical produced in New England was Puritan rum!!

New England was also filled with shrubs and flowers. Many of these plants were not known in Europe, and were a good source of flavors, fragrances and pharmaceuticals. John Winthrop, Jr., and many other Colonial physicians explored the world of extracts and dried herbs. Medicinal chemistry thrived in New England, both then and now. In addition to plant based medicines, John Winthrop, Jr., also produced remedies based on animal and mineral sources. His most famous pill, *rubila*, was based on antimony trioxide.

Since there were very few members of the Royal College of Physicians in Colonial New England, iatrochemists were free to develop their craft without arrest or harassment. The success of this venture among the Congregational clergy is one of the best stories from this era. They were highly educated, called to ministry, and served all the needs of their parish: spiritual, physical and emotional. They were also underpaid, and practicing medicine helped them to make ends meet. They also needed to farm, and their wives needed an additional craft business. St. Paul understood the perils of preaching "for a living."

After John Winthrop, Jr., died, his son, Waitstill, continued in his alchemical pursuits. But, he was not an intellectual, did not have a worldwide circle of alchemists, and did not generate any new ideas in chemistry. As an isolated Colonial, his thoughts stagnated and he produced no disciples, unlike his father. John Winthrop, Jr., gave a fine telescope to Harvard College. It was used regularly by later Harvard faculty and astronomy replaced alchemy as the favorite topic of Harvard's Natural Philosophy. Not until the nineteenth century did Harvard produce a world-class Chemistry Professor, Josiah Parsons Cooke (1827–1894), the Erving Professor of Chemistry. While he knew no chemistry at the time of his appointment, and only spent a few months in Paris with Dumas and Regnault, he read everything he could get his hands on, carried out detailed experiments, and developed a truly coherent course in chemistry [6].

Since alchemy was thriving in Connecticut in the seventeenth century, the question of its demise upon the death of John Winthrop, Jr., must be addressed. One major problem was that alchemy is expensive! Since gold was never produced from anything else, and the alkahest does not exist at all, money was constantly wasted on fruitless experiments. Even as famous an alchemist as George Starkey spent considerable time in debtor's prison [7]. Pursuing alchemy required a reliable source of funds. Patrons soon tired of the negative results. Medicine did bring in money for some iatrochemists that allowed them to waste it on alchemy. Other alchemists owned commercial enterprises that fed their addiction. John Winthrop, Jr., owned land, agricultural and husbandry businesses, governorships, etc. so that he could keep the wolf from the door. But, he died in serious debt.

While the practical chemistry of Johann Glauber kept manufacturing chemists busy for the seventeenth century [8], the theoretical basis for chemistry was not strong enough to produce growth. The notion that all substances were composed of mercury, sulfur and salt did not lead to new discoveries that authenticated this paradigm. That did not stop some chemists from pursuing this approach well into the nineteenth century. I believe that the key development that turned chemists away from alchemy was the advent of Newtonian Natural Philosophy. This is rich, considering that Isaac Newton was a serious alchemist [9]. But his secret experiments in the arcana were not widely known, and his publications in his *Opticks* suggested that matter was composed of microscopic particles interacting by occult forces [10].

Two of the leading chemists of the eighteenth century, Hermann Boerhaave (1669–1738) [11] and Roger Boscovich (1711–1787) [12] were Newtonians. While this paradigm was elegant, and the so-called Boscovich potential is baroque, it did not lead to a coherent view of chemistry that stimulated the sort of experiments that could revolutionize a science. Boerhaave was still operating largely along alchemical lines in the laboratory. For example, he distilled a single sample of mercury more than 500 times [13]. His monumental text, *Elementa Chemiae* (1732), is grounded in hundreds of well-described experiments, and contains a good descriptive chemistry of the known elements. Other leading chemists of the eighteenth century were fascinated by fictive substances like phlogiston. Even as great a chemist as Joseph Priestley (1733–1804) held to the existence and importance of a fictive substance of negative weight that pervaded metals when they were burned. While many good chemical things did happen in the eighteenth century, like the formulation of stoichiometry, no real progress was likely until a central paradigm for chemistry was proposed: the real existence of chemical atoms.

The basis for a successful world of chemistry in seventeenth century New England was entirely empirical. Certain recipes did produce the desired products, even if the chemists were largely ignorant of the real reasons why. Saltpeter is a prime example. Elite natural philosophers like Benjamin Worsley [14] were clueless about the actual processes involved, while artisanal "petermen" knew how to taste the "nitre-bed" to determine if potassium nitrate was present. John Winthrop, Jr., read and considered the arcana of classical alchemy, but he followed recipes when it came to producing the needed chemicals.

The rise of Chemical Philosophy in the sixteenth and seventeenth centuries was a noble exercise [15], but when the underlying axioms are actually false, no amount of "good arguments" will lead to the truth. Nevertheless, careful observation, detailed laboratory notebooks, and honest reporting did advance the craft of chemistry. A sound natural philosophy is only as good as its empirical foundation. Eventually, consistent results in the analysis of chemical reactions revealed a law of definite proportions. Careful measurements of mass and volume, combined with reasonably pure chemicals, led to good estimates for the elemental composition of many minerals and natural products. Measurements of the heat associated with chemical processes added another layer of sophistication. But, all of this more modern chemistry was lacking in seventeenth century New England. It would not arrive until around 1850!

References

1. Kurlansky M (2002) Salt: a world history. Penguin Books, New York
2. Evelyn J (1664) Sylva, or, a discourse of forest trees and the propagation of timber in his majesties dominions. Royal Society, London
3. Cressy D (2013) Saltpeter: the mother of gunpowder. Oxford University Press, Oxford
4. National Park Service (1941) American charcoal making in the era of the cold-blast furnace. United States Department of the Interior
5. Hartley EN (1957) Ironworks on the Saugus. University of Oklahoma Press, Norman
6. Cooke JP (1870) First principles of chemical philosophy. Sever, Francis and Co., Boston
7. Newman WR (2003) Gehennical fire: the lives of George Starkey. University of Chicago Press, Chicago
8. Glauber J (1689) The works of the highly experienced and famous chymist, John Rudolph Glauber, London
9. White M (1997) Isaac Newton: the last sorcerer. Addison-Wesley, Reading
10. Buchwald JZ, Cohen IB (2001) Isaac Newton's natural philosophy. MIT Press, Cambridge
11. Powers JC (2012) Inventing chemistry: Herman Boerhaave and the reform of the chemical arts. University of Chicago Press, Chicago
12. Whyte LW (1961) Roger Joseph Boscovich, S.J., F.R.S., 1711–1787: studies of his life and work on the 250th anniversary of his birth. Fordham University Press, New York
13. Boerhaave H (1734) Some experiments concerning mercury. Royal Society, London
14. Leng T (2008) Benjamin Worsley (1618–1677): trade, interest and the spirit in revolutionary England. The Royal Historical Society
15. Debus AG (1977) The chemical philosophy: Paracelsian science and medicine in the sixteenth and seventeenth centuries. Science History Publications, New York

Printed in the United States
By Bookmasters